Springer Tracts in Nature-Inspired Computing

The book series is aimed at providing an exchange platform for researchers to summarize the latest research and developments related to nature-inspired computing in the most general sense. It includes analysis of nature-inspired algorithms and techniques, inspiration from natural and biological systems, computational mechanisms and models that imitate them in various fields, and the applications to solve real-world problems in different disciplines. The book series addresses the most recent innovations and developments in nature-inspired computation, algorithms, models and methods, implementation, tools, architectures, frameworks, structures, applications associated with bio-inspired methodologies and other relevant areas.

The book series covers the topics and fields of Nature-Inspired Computing, Bio-inspired Methods, Swarm Intelligence, Computational Intelligence, Evolutionary Computation, Nature-Inspired Algorithms, Neural Computing, Data Mining, Artificial Intelligence, Machine Learning, Theoretical Foundations and Analysis, and Multi-Agent Systems. In addition, case studies, implementation of methods and algorithms as well as applications in a diverse range of areas such as Bioinformatics, Big Data, Computer Science, Signal and Image Processing, Computer Vision, Biomedical and Health Science, Business Planning, Vehicle Routing and others are also an important part of this book series.

The series publishes monographs, edited volumes and selected proceedings.

More information about this series at http://www.springer.com/series/16134

Simon James Fong · Richard C. Millham
Editors

Bio-inspired Algorithms for Data Streaming and Visualization, Big Data Management, and Fog Computing

Springer

Editors
Simon James Fong
University of Macau
Taipa, China

Richard C. Millham
Durban University of Technology
Durban, South Africa

ISSN 2524-552X ISSN 2524-5538 (electronic)
Springer Tracts in Nature-Inspired Computing
ISBN 978-981-15-6694-3 ISBN 978-981-15-6695-0 (eBook)
https://doi.org/10.1007/978-981-15-6695-0

This Springer imprint is published by the registered company Springer Nature Singapore Pte Ltd.
The registered company address is: 152 Beach Road, #21-01/04 Gateway East, Singapore 189721, Singapore

Preface

The purpose of this book is to provide some insights into recently developed bio-inspired algorithms within recent emerging trends of fog computing, sentiment analysis, and data streaming as well as to provide a more comprehensive approach to the big data management from pre-processing to analytics to visualisation phases. Although the application domains of these new algorithms may be mentioned, these algorithms are not confined to any particular application domain. Instead, these algorithms provide an update into emerging research areas such as data streaming, fog computing, and phases of big data management.

This book begins with the description of bio-inspired algorithms with a description on how they are developed, along with an applied focus on how they can be applied to missing value extrapolation (an area of big data pre-processing). The book proceeds to chapters including identifying features through deep learning, overview of data mining, recognising association rules, data streaming, data visualisation, business intelligence and current big data tools.

One of the reasons for writing this book is that the bio-inspired approach does not receive much attention although it continues to show considerable promise and diversity in terms of approach of many issues in big data and streaming. This book outlines the use of these algorithms to all phases of data management, not just a specific phase such as data mining or business intelligence. Most chapters demonstrate the effectiveness of a selected bio-inspired algorithm by experimental evaluation of it against comparative algorithms. One chapter provides an overview and evaluation of traditional algorithms, both sequential and parallel, for use in data mining. This chapter is complemented by another chapter that uses a bio-inspired algorithm for data mining in order to enable the reader to choose the most appropriate choice of algorithms for data mining within a particular context. In all chapters, references for further reading are provided, and in selected chapters, we will also include ideas for future research.

Taipa, China — Simon James Fong
Durban, South Africa — Richard C. Millham

Contents

About the Editors

Simon James Fong graduated from La Trobe University, Australia, with a First-Class Honours B.E. Computer Systems degree and a Ph.D. Computer Science degree in 1993 and 1998, respectively. Simon is now working as an Associate Professor at the Computer and Information Science Department of the University of Macau. He is a Co-Founder of the Data Analytics and Collaborative Computing Research Group in the Faculty of Science and Technology. Prior to his academic career, Simon took up various managerial and technical posts, such as Systems Engineer, IT Consultant, and E-commerce Director in Australia and Asia. Dr. Fong has published over 500 international conference and peer-reviewed journal papers, mostly in the areas of data mining, data stream mining, big data analytics, meta-heuristics optimization algorithms, and their applications. He serves on the editorial boards of the Journal of Network and Computer Applications of Elsevier, IEEE IT Professional Magazine, and various special issues of SCIE-indexed journals. Currently, Simon is chairing a SIG, namely Blockchain for e-Health at IEEE Communication Society.

Richard C. Millham a B.A. (Hons.) from the University of Saskatchewan in Canada, M.Sc. from the University of Abertay in Dundee, Scotland, and a Ph.D. from De Montfort University in Leicester, England. After working in industry in diverse fields for 15 years, he joined academe and he has taught in Scotland, Ghana, South Sudan, and the Bahamas before joining DUT. His research interests include software and data evolution, cloud computing, big data, bio-inspired algorithms, and aspects of IOT.

Chapter 1
The Big Data Approach Using Bio-Inspired Algorithms: Data Imputation

Richard Millham, Israel Edem Agbehadji, and Hongji Yang

1 Introduction

In this chapter, the concept of big data is defined based on the five characteristics namely velocity, volume, value, veracity, and variety. Once defined, the sequential phases of big data are denoted, namely data cleansing, data mining, and visualization. Each phase consists of several sub-phases or steps. These steps are briefly described. In order to manipulate data, a number of methods may be employed. In this chapter, we look at an approach for data imputation or the extrapolation of missing values in data. The concept of genetic algorithms along with its off-shoot, meta-heuristic algorithms, is presented. A specialized type of meta-heuristic algorithm, bio-inspired algorithms, is introduced with several example algorithms. An example, a bio-inspired algorithm, the kestrel, is introduced using the steps outlined for the development of a bio-inspired algorithm (Zang et al. 2010). This kestrel algorithm will be used as an approach for data imputation within the big data phases framework.

R. Millham (✉) · I. E. Agbehadji
ICT and Society Research Group, Durban University of Technology, Durban, South Africa
e-mail: richardm1@dut.ac.za

I. E. Agbehadji
e-mail: israeldel2006@gmail.com

H. Yang
Department of Informatics, University of Leicester, Leicester, England, UK
e-mail: Hongji.Yang@Leicester.ac.uk

S. J. Fong and R. C. Millham (eds.), *Bio-inspired Algorithms for Data Streaming and Visualization, Big Data Management, and Fog Computing*,
Springer Tracts in Nature-Inspired Computing,
https://doi.org/10.1007/978-981-15-6695-0_1

2 Big Data Framework

The definition of big data varies from one author to another. A common definition might be that it denotes huge volume and complicated data sets because it comes from heterogeneous sources (Banupriya and Vijayadeepa 2015). Because of the enormous variety in definitions, big data is often known by its characteristics of velocity, volume, value, veracity, and variety which constitutes the framework of big data. Velocity relates to how quickly incoming data needs to be evaluated with results produced Longbottom and Bamforth (2013). Volume relates to the amount of data to be processed. Veracity relates to the accuracy of results emerging from the big data processes. Value is the degree of worth that the user will obtain from the big data analysis.

3 Evolutionary and Bio-Inspired Methods

Genetic algorithms (GA) inherited the principles of "Darwin's Evolutionary Theory". Genetic algorithms provide solutions to a search problem by using biological evolution principles. Nature breeds a large number of optimized solutions which have been discovered and deployed to solve problems (Zang et al. 2010). Genetic algorithm adopts some common genetic expressions such as

(1) **Chromosome**: where the solution to an optimization problem is encoded (Cha and Tappert 2009).
(2) **Selection**: a phase where individual chromosomes are evaluated and the best are chosen to raise the next generation.
(3) "**Crossover**" and "mutation" are genetic methods for pairing parents to change their genetic makeup through the process of breeding.

The first phase of a genetic algorithm produces initial population which represents randomly generated individuals. This individual forms a range of potential solutions in which the population size is determined by the nature of the problem. The initial population represents the search space and the algorithm begins with an initial estimate. Then the operators of crossover and mutation are applied to the population in order to try to improve the estimate through evolution (Agbehadji 2011). The next phase assesses the individual of a given population to determine their fitness values through a fitness function. The higher the fitness value of individuals, the greater the probability that the individual will be selected for the next generation. This process of mutation, selection via the fitness function, and generation/iteration continues until a termination criteria, or final optimal value or solution, is met.

Because of its adaptive and iterative nature, genetic algorithm may be used to discover multiple types of optimal solutions from a set of initial random potential solutions. Given the continuous updating of the population through the application of genetic operators and the culling-off weak generation via a fitness function, the gradual improvement of the population to a termination condition, or optimal solution, is made. One such solution that may be determined via genetic algorithm is discovering an optimal path (Dorigo et al. 2006). A practical application of a genetic algorithm is extrapolating missing values in a dataset (Abdella and Marwala 2006).

Meta-heuristic search or bio-inspired search or nature-inspired search methods are mostly used interchangeably to refer to search algorithms developed from the behaviour of living creatures in their natural habitat. Conceptually, living creatures are adapted to make random decisions that can steer them either towards hunt or away from its enemy. Meta-heuristic search methods can be combined to develop a more robust search algorithm for any complex problems. The advantage of meta-heuristic search method is the ability to ignore a search that is not promising. Generally, meta-heuristic search algorithm begin with random set of individuals where each represents a possible solution. In each generation, instead of mutation, there is a random Levy walk (which corresponds to the random movements of animals/random searches for an optimal solution). At the end of each generation, the fitness of each individual of that generation is evaluated via a specified fitness function. Only those individuals that meet a prescribed threshold of fitness, as determined by a fitness function, are allowed to continue as parents for the next generation. The succession of generation continues until some pre-defined stopping criteria is reached; ideally, this stopping criteria is when a near-optimal solution has been found (Fong 2016).

3.1 Development Process for Bio-Inspired Algorithms

These are the stages in developing a bio-inspired algorithms:

(a) Firstly: identify the unique behaviour of a creature in nature,

(b) Secondly: formulate basic expressions on their behaviour.

(c) Thirdly, transform the basic expression into mathematical equation, identify some underlying assumptions, and setup initial parameters,

(d) Fourthly, write a pseudo-code to represent the basic expression,

(e) Fifth: test the code on actual data and refine the initial parameter for better performance of the algorithm.

Usually, animal behaviour constitutes actions relative to its environment and context; thus, a particular animal behaviour should be modelled in conjunction with other animal behaviours, other in terms of a team of individuals or another species, in order to achieve better results. Therefore, the nature-inspired algorithms can be combined with other algorithms for an efficient result and more robust algorithm (Zang et al. 2010).

3.1.1 Examples of Bio-Inspired Algorithms

Bio-inspired algorithms can focus on the collective behaviour of multiple simple individuals (as in particle swarm) (Selvaraj et al. 2014), the co-operative behaviour of more complex individuals (as in wolf search algorithm) (Tang et al. 2012), or the single behaviour of an individual (Agbehadji et al. 2016b). Within these categories, such as particle swarm, there are many types (such as artificial bee colony), and within these types, there are many applications of the same algorithm for such things as image processing, route optimization, etc. (Selvaraj et al. 2014).

A major category of bio-inspired algorithms are particle swarm algorithms. Particle swarm algorithms is a bio-inspired technique that mimics the swarm behaviour of animals such as fish schools or bird flocks (Kennedy and Eberhart 1995). The behaviour of the swarm is determined by how particles adapt and make decisions in changing their position within a space relative to the positions of neighbouring particles. The advantage of swarm behaviour is that as particles make a decision, it leads to local interaction among particles which in turn, lead it to an emergent behaviour (Krause et al. 2013). Particle swarm algorithm that focuses on finding the near-optimal solution includes the firefly algorithm, bats (Yang and Deb 2009) and cuckoo birds (Yang and Deb 2009).

3.1.2 Firefly Algorithm

The basis of the firefly algorithm's behaviour is the short and rhythmic flashes it produces. This flashing light of fireflies is used as an instrument to attract possible prey, attract mating partners, and to act as a warning signal. The firefly signalling system consists of rhythmic flash, frequency of flashing light and time period of flashing. This signalling system is controlled by simplified basic rules underlining the behaviour of firefly that can be summarized as, one firefly can be connected with another; hence, this connection which refers to attractiveness is proportional to the level of brightness between each firefly and brightness is affected by landscape (Yang 2010a, b, c). The attraction formulation is based on the following assumptions:

(a) Each firefly attracts another fireflies that has a weak flash light
(b) This attraction depends on the level of brightness of the flash which is reversely proportional to their proximity to each other
(c) The firefly with the brightest flash is not attracted to any other firefly and their flight is random (Yang 2010a, b, c).

The signal of this flashing light instrument is governed by a simplified basic rule which forms the basis of firefly behaviour. In comparison with a genetic algorithm, it uses what is referred to as operators that are mutation, crossover, and selection. The firefly uses attractiveness and brightness of its flashing light. The similarity between the firefly algorithm and the genetic algorithm is that both algorithms generate an initial population which is updated continuously at each iteration, via fitness function. In terms of firefly behaviour, the brighter fireflies attract those fireflies nearest to them and those fireflies whose brightness fall below a defined threshold are removed from subsequent population. The brightest fireflies, whose brightness have exceeded a specified threshold, constitute the next generation and this generation continues until either a termination criteria (best possible solution) is met or the highest number of iterations is achieved. The use of brightness in firefly algorithm is to help attract the weaker firefly which mimics the extrapolation of missing values in a dataset where the fireflies represent known values and those with the brightest light (indicating closeness to the missing values as well as nearness to the set of data including the missing value) are selected as suitable to replace the missing value entries.

3.1.3 Bat Search Algorithms

The bat search algorithm is another bio-inspired search technique that is grounded on the behaviour of micro-bats within their natural environment (Yang 2010a, b, c). Bat is known to have a very unique behaviour called echolocation. This characteristic assists bats to orient themselves and find prey within their habitat. The search strategy of a bat, whether to navigate or to capture prey, is governed by the pulse rate and loudness of their cry. This pulse rate governs the enhancement of the best possible solution, its loudness affects the acceptance of the best possible solution (Fister et al. 2014). Similar to genetic search algorithm, the bat search algorithm begins with random initialization, evaluation of the newly generated population, and after multiple iterations, the best possible solution is outputted. In contrast to the wolf search algorithm that uses attractiveness, the bat search algorithm uses its pulse rate and loudness to steer its search for a near-optimal solution. The bat search algorithm, with its behaviour, has been applied to several optimization problems to find the best possible solution.

3.1.4 Wolf Search Algorithm

The wolf search algorithm (WSA) is a nature-inspired algorithm that focuses on a wolf's preying behaviour (Tang et al. 2012). This preying behaviour, as derived from wolves' behaviour, demonstrates that wolves are able to hunt independently by recalling their own trait; have the ability to join with a fellow wolf only when the other wolf is in a better position; and have the ability to randomly escape when a hunter appears. This expressed wolf behaviour allows them to adapt to their habitat when hunting for prey. Because wolves have the ability to join a fellow wolf in a better position, it implies that wolves have some trust in each other and they avoid preying upon each other. In addition, wolves prefer to only move into territory mark by other wolves which indicates that the territory is safe for other wolves to live in. Moreover, if this new location is better, the motivation is stronger especially if this new location is within territory already occupied by a fellow wolf. This wolf search algorithm can be defined as a search procedure which begins with setting the initial population, evaluating the candidate population and updating the current population via fitness test, and continuous until stopping criteria is met. Particle swarm algorithms, like firefly, attract its prey by using the characteristics of attractiveness and brightness while wolf uses the characteristic of attractiveness of prey within its visual range. Wolves also have both individual search capability and independent flocking movement. In WSA, consequently, the swarming behaviour of wolves, unlike other nature-inspired algorithms, is delegated to individual wolf instead of a single leader, as is the case in the particle swarm and firefly algorithms. In practice, WSA works as if there are "multiple leaders swarming from multiple directions" to the best possible solution instead of a "single flock" that searches for the best possible solution in one direction at a time (Tang et al. 2012). Similar to the firefly and bat, the WSA characteristic and behaviour towards attraction can be used to extrapolate the estimated value that is near to known values in any missing data imputation method.

Nature-inspired or bio-inspired search algorithms are characterized by randomization, efficient local search and finding of global best results (Yang 2010a, b, c). With the newly developed kestrel search algorithm, the random circling of a kestrel is examined to see how it may be used to achieve a best possible solution (estimates closest to missing values). The advantage of the random encircling method of the kestrel, unlike other bio-inspired algorithms, is that it maximizes the local search space, and in so doing, it creates a wider range of possible solutions, based on a hovering position, in order to assess and obtain the best possible solution.

3.1.5 Kestrel Behaviour

In keeping with Zang et al. (2010)'s prescribed method of developing a bio-inspired search algorithm, in this case that of a kestrel bird, the behaviour is observed and briefly summarized to depict its behaviour in a natural environment. This search algorithm of a kestrel bird is based on its hunting characteristics that are either hovering or perched hunt. Kestrels are highly territorial and hunt individually (Shrubb 1982; Varland 1991). One researcher, Varland (1991), recognized that during hunts, kestrels tend to be imitative rather than co-operative. In other words, kestrels choose "not to communicate with each other" instead they "imitate the behaviour of other kestrels with better hunting techniques". With mimicking better techniques comes the improvement of their own technique. The hunt technique, however, can be dependent on such factors such as the type of prey, current weather conditions, and energy requirements (for gliding or diving) (Vlachos et al. 2003).

During their hunt, kestrels use their "eyesight to view small and agile prey" within its coverage area, as defined by its visual circling radius. Prey availability is indicated either through a "trail of urine and faeces" from ground-based prey or through the minute air disturbance from airborne-based prey. Once the prey availability is identified, the "kestrel positions itself to hunt". Kestrels can "hover in changing airstream, maintain fixed forward" looking position with its eye on a potential prey, and use "random bobbing" of its head to find the minimum distance between its "position and the position of prey". Because kestrels can view ultraviolet light, they are able to discover trails of urine and faeces left by prey such as voles (Honkavaara et al. 2002).

While hovering, kestrels can perform a broader search (e.g. global exploration) across different territories within its "visual circling radius", are able to "maintain a motionless position with their forward-looking eyes fixed on prey, detect tiny air disturbances from flying prey (especially flying insects) as indicators of prey", and can move "with precision through a changing airstream". Kestrels have the ability to flap their winds and adjust their long tails in order to stay in a place (denoted as a still position) in a "changing airstream". While in perch mode (often perching from high fixed structures such as poles or trees), kestrels change their perch position every few minutes before performing a thorough search (which is denoted as "local exploitation" based on its individual hunt behaviour) of its local territory which requires "less energy than a hovering hunt". While in perch mode, the kestrel uses its ultraviolet detection capacity to discover potential prey such as voles nearer to its perch area. This behaviour suggests that while in perch stance, kestrel uses this position to conserve some energy and to focus their ultraviolet detection capabilities for spotting slow moving prey on the ground. Regardless of perch or hovering mode, skill development also plays a role. Individual kestrels with better "perch and hovering skills" that are utilized in a larger search area possess a better chance to swoop down faster on their prey or flee from its enemies than "individual kestrels that develop hunting skills in local territories" (Varland 1991). Consequently, it is important to combine hunting skills from both hovering and perch modes in order to accomplish a successful hunt.

In order to better characterize the kestrel, certain traits are given as their defining behaviour:

(1) Soaring: it provides a wider search space (global exploration) within their visual coverage area
    (a) Still (motionless) location with eyesight set on prey
        (i) Encircles prey underneath it using its keen eyesight
(2) Perching: this enables thorough search or local exploitation within a visual coverage radius
    (a) Behaviour involves "frequent bobbing of head" to find the best position of attack
    (b) Using a trail, identify potential prey and then the kestrel glides to capture prey

These behavioural characteristics are based on the following assumptions:

(a) The still position of the kestrel bird provides a near perfect circle. Consequently, frequent changes in circle direction depend on the position of prey shifting the centre of this circling direction
(b) The frequent bobbing of the kestrel's head provides a "degree of magnified or binocular vision" that assists in judging the distance from the kestrel to a potential prey and calculating a striking move with the required speed
(c) "Attractiveness is proportional to light reflection". Consequently, "the higher or longer a distance from the trail to the kestrel, the less bright of a trail". This distance parameter applies to both the hovering height and the distance away from the perch.

(d) "New trails are more attractive than old trails". Thus, the trail decay, as the trail evaporates, depends on "the half-life of the trail".

Mathematical Model of Kestrel's Behaviour

Following the steps of Zang et al. (2010), a model that represents the kestrel behaviour is expressed mathematically. The following sets of kestrel characteristics, with their mathematical equivalents, are provided below:

- **Encircling behaviour**

This encircling behaviour occurs when the "kestrel randomly shifts (or changes)" its "centre of circling direction" in response to detecting the current position of prey. When the prey changes from its present position, the kestrel randomly shifts, or changes, the "centre of circling direction" in order to recognize the present position of prey. With the change of position of prey, the kestrel correspondingly alters its encircling behaviour to encircle its prey. The movement of prey results in the kestrel adopting the best possible position to strike. This encircling behaviour $\vec{D}$ (Kumar 2015) is denoted in Eq. 1 as:

$$\vec{D} = \left| \vec{C} * \overrightarrow{x_p}(t) - \vec{x}(t) \right| \tag{1}$$

$\vec{C}$is denoted in equation 2as:

$$\vec{C} = 2 * \overrightarrow{r_1} \tag{2}$$

where $\vec{C}$ is the "coefficient vector", $\overrightarrow{x_p}(t)$ is the position vector of the prey, and $\vec{x}(t)$ represents the position of a kestrel, r_1 and r_2 are random values between 0 and 1 indicating random movements.

- **Current position**

The present best position of the kestrel is denoted in Eq. 3 as follows:

$$\vec{x}(t+1) = \overrightarrow{x_p}(t) - \vec{A} * \vec{D} \tag{3}$$

Consequently, the coefficient $\vec{A}$ is denoted in Eq. 4 as follows:

$$\vec{A} = 2 * \vec{z} * \overrightarrow{r_2} - \vec{z} \tag{4}$$

where $\vec{A}$ also represents coefficient vector, $\vec{D}$ is the encircling value acquired, $\overrightarrow{x_p}(t)$ is the prey's position vector, $\vec{x}(t+1)$ signifies present best position of kestrels. $\vec{z}$ decreases linearly from 2 to 0 and this value is also used to "control the randomness" at each iterations. The $\vec{z}$ is denoted in Eq. 5 as follows:

$$\vec{z} = \vec{z}_{hi} - (\vec{z}_{hi} - \vec{z}_{\text{low}})\frac{\text{itr}}{\text{Max\_ itr}} \tag{5}$$

where *itr* is the current iteration, *Max_itr* represents maximum number of iterations that stop the search, z_{hi} denotes the higher bound of 2, z_{low}denotes the lower bound of 0. Any other kestrels included in this search for prey will update their position based on the best position of the leading kestrel. In addition, the change in position in the airstream for kestrels is dependent on the "frequency of bobbing", how it attracts prey and "trail evaporation". These dependent variables are denoted as follows:

(a) **Frequency of bobbing**

The bobbing frequency is used to determine sight distance measurement within the search space. This is denoted in Eq. 6 as follows:

$$f_{t+1}^{k} = f_{\text{min}} + (f_{\text{max}} - f_{\text{min}}) * \alpha \tag{6}$$

where $\alpha \in [0, 1]$ indicates a random number to govern the "frequency of bobbing within a visual range". The maximum frequency f_{max} is set at 1 while the minimum frequency f_{min} is set at 0.

(b) **Attractiveness**

Attractiveness β denotes the light reflection from trails, which is expressed in Eq. (7) as follows:

$$\beta(r) = \beta_o e^{-\gamma r^2} \tag{7}$$

where β_o equals l_o and constitutes the initial attractiveness, γ denotes variation of light intensity between [0, 1]. r denotes the sight distance $s(x_i, x_c)$ measurement which is calculated using "Minkowski distance" expression in Eq. (8) as:

$$s(x_i, x_c) = (\sum_{k=1}^{n} |x_{i,k} - x_{c,k}|^{\lambda})^{\frac{1}{\lambda}} \tag{8}$$

Consequently, Eq. 9 expresses the visual range as follows:

$$V \leq s(x_i, x_c) \tag{9}$$

where x_i denotes the current sight measurement, x_c indicates all possible adjacent sight measurement near x_i, n is the total number of adjacent sights and λ is the order (values of 1 or 2) and V is the visual range.

(c) **Trail evaporation**

A trail may be defined as way to form and maintain a line (Dorigo and Gambardella 1997). In meta-heuristic algorithms, trails are used by ants to track the path from their home to a food source while avoiding getting mired to just one food source. Thus, these trails enable multiple food sources to be used within a search space. (Agbehadji 2011) While ants search continuously, trails are developed with elements attached to these trails. These elements assist ants in communicating with each other regarding the position of food sources. Consequently, other ants constantly follow this path while depositing elements for the trail to remain fresh. In the same manner that ants use trails, "kestrels use trails to search for food sources". These trails, unlike those of ants, are created by prey which, thus, provide an indication to kestrels on the obtainability of food sources. The assumption with the kestrel is that the elements left by these prey (urine, faeces, etc.) are similar to those elements left on an ant trail. In addition, when the food source indicated by the trail is exhausted, kestrels no longer pursue this path as the trail elements begin to reduce with "time at an exponential rate". With the reduction of trails' elements, the trail turns old. This reduction indicates the unstable essence of trail elements which is expressed as if there are N "unstable substances" with an "exponential decay rate" of γ, then the equation to detail how N element reduces in time t is expressed as follows (Spencer 2002):

$$\frac{\mathrm{d}N}{\mathrm{d}t} = -\gamma N \tag{10}$$

Because these elements are unstable, there is "randomness in the decay process". Consequently, the rate of decay (γ) with respect to time (t) can be re-defined as follows:

$$\gamma_t = \gamma_o e^{-\lambda t} \tag{11}$$

where γ_o is a "random initial value" of trail elements that is reduced at each iteration. t is the number of iterations/generations/time steps, where $t \in [0, Max\_itr]$ with Max_itr being the maximum number of iterations.

$$\text{if } \gamma_t \rightarrow \begin{cases} \gamma_t > 1, & \text{trail is new} \\ 0, & \text{otherwise} \end{cases} \tag{12}$$

Once more, the decay constant λ is denoted by:

$$\lambda = \frac{\phi\text{max} - \phi\text{min}}{t_{\frac{1}{2}}} \tag{13}$$

where λ is "the decay constant", ϕmax is the maximum number elements in trail, ϕmin is the minimum number of elements in trail and $t_{\frac{1}{2}}$ is the "half-life period of a trail which indicates that a trail" has become "old and unattractive" for pursuing prey.

Lastly, the Kestrel will updates its location using the following equation:

$$x_{t+1}^k = x_t^k + \beta_o e^{-\gamma r^2}\left(x_j - x_i\right) + f_t^k \tag{14}$$

where x_{t+1}^k signifies the present optimal location of kestrels. x_t^k is the preceding location.

- **Fitness function**

In order to evaluation how well an algorithm achieves in terms of some criteria (such as the quality of estimation for missing value), a fitness function is applied. In the case of missing value estimation, the measurement of this achievement is in terms of "minimizing the deviation of data points from the estimated value". A number of performance measurement tools may be used such as mean absolute error (MAE), root mean square (RMSE), and mean square error (MSE).

In this chapter, the fitness function for the kestrel search algorithm uses the mean absolute error (MAE) as its performance measurement tool in order to determine the quality of estimation of missing values. MAE was selected for use in the fitness function because it allows the modelled behaviour of the kestrel to fine tune and improve on its much more precise estimation of values concern for negative values. The MAE is expressed in Eq. (15) as follows:

$$\text{MAE} = \frac{1}{n}\sum_{i=1}^{n}|o_i - x_i| \tag{15}$$

where x_i indicates the estimated value at the ith position in the dataset, o_i denotes the observed data point at ith position "in the sampled dataset, and n is the number of data points in the sampled dataset".

- **Velocity**

The velocity of kestrel as it moves from its current optimal location in a "changing airstream" is expressed as:

$$v_{t+1}^k = v_t^k + x_t^k \tag{16}$$

Any variation in velocity is governed by the inertia weight ω (which is also denoted as the convergent parameter). This "inertia weight has a linearly" diminishing value. Thus, velocity is denoted in Eq. 17 as follows:

$$v_{t+1}^k = \omega v_t^k + x_t^k \tag{17}$$

where ω is the "convergence parameter", v_t^k is the "initial velocity", x_t^k is best location of the kestrel and the v_{t+1}^k is the present best velocity of the kestrel. Kestrels explore through the search space to discover optimal solution and in so doing, they constantly update the velocity, random encircling, and location towards the best estimated solution.

Table 1 Kestrel algorithm

- Set parameters
- Initialize population of *n* Kestrels using equation (3) and evaluate fitness of population using equation (18)
- Start iteration (loop until termination criteria is met)
  - Compute Half-life of trail using equation (11)
  - Compute frequency of bobbing using equation (6)
  - Evaluate position for each Kestrel as in equation using equation (14)
    - If f $(x_i) <$ f(x_j) then
      - Move Kestrel *i* towards *j*
    - End if
- Update position f(x_i) for all *i*=1 to n as in equation (17)
- Find the current best value
- End loop

Kestrel-Based Search Algorithm

Following Zang (2010) steps to develop a new bio-inspired algorithm, after certain aspects of behaviour of the selected animal is mathematically modelled, the pseudo-code or algorithm that incorporates parts of this mathematical model is developed both to simulate animal behaviour and to discover the best possible solution to a given problem.

The algorithm for kestrel is given as follows (Table 1).

Implementation of Kestrel-Based Algorithm

After the algorithm for the newly developed bio-inspired algorithm has been determined, the next step, according to Zang et al. (2010) is to test the algorithm experimentally. Although kestrel behaviour, due to its encircling behaviour and adaptability to different hunting contexts [either high above as in hovering or near the ground as in perching] (Agbehadji et al. 2016a), is capable of being used in a variety of steps and phases of big data mining, the step of estimating missing values within the data cleansing phase was chosen.

Following Zang's et al. (2010) prescription to develop a bio-inspired algorithm, the parameters of the bio-inspired algorithm are set. The initial parameters for the KSA algorithm were set as $\beta_o = 1$ with visual range = 1. As per Eq. 5, the parameters for the lower and higher bound, $z_{\min} = 0.2$ and $z_{\max} = 0.9$, respectively, were set accordingly. A maximum number of 500 iterations/generations were set in order to allow the algorithm to have a better opportunity of further refining the best estimated values in each iteration.

Further to Zang's et al. (2010) rule, the algorithm is tested against appropriate data. This algorithm was tested using a representative dataset matrix of 46 rows and 9 columns with multiple values missing in each row of the matrix. This matrix was designed to allow for a thorough testing of estimation of missing values by the KSA

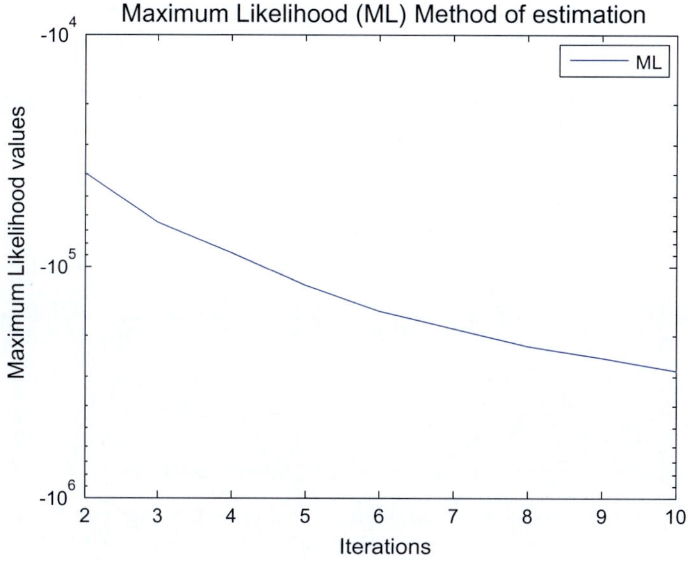

Fig. 1 Maximum likelihood

algorithm. This testing produced the following Fig. 1: A "sample set of data (46 by 9 matrix) with multiple missing values in the row matrix was used in order to provide a thorough test of missing values in each row of a matrix". The test revealed the following figure represented as Fig. 2:

Figure 2 shows a single graph of the fitness function value of the KSA algorithm during "500 iterations". As can be seen in this graph, the "curve ascends and descends steeply during the beginning iterations and then gradually converges at the best possible solution at the end of 500 iterations/generations". The steps within the curve symbolize looking for a best solution within a particular search space, using a random method, until one is found and then another space is explored. The curve characteristics indicate that at the starting iterations, the KSA algorithm "quickly maximizes the search space and then gradually minimizes" until it converges to the best possible optimal value.

4 Conventional Data Imputation Methods

Conventional approaches to estimate missing data values include ignoring missing attributes or fill in missing values with a global constant (Quinlan 1989), with the real possibility of detracting from the quality of pattern(s) discovered based on these values. Based on the historical trend model, missing data may be extrapolated, in terms of their approximate value, using trends (Narang 2013). This procedure is

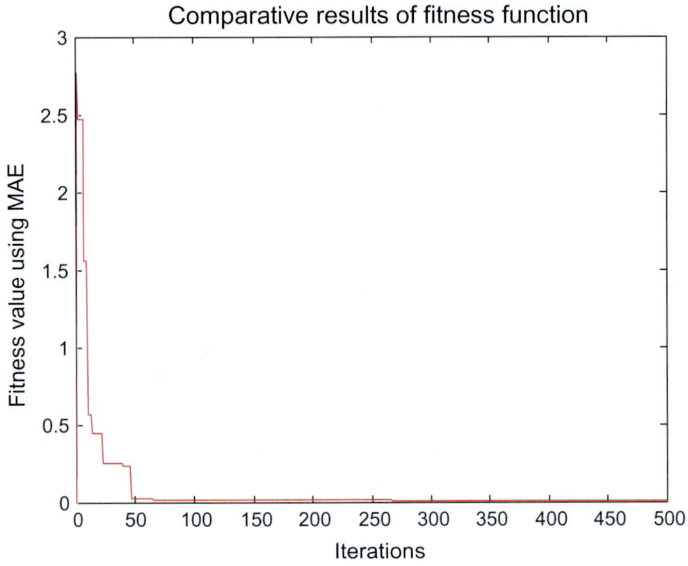

Fig. 2 KSA fitness

common in the domain of real-time stock trading with missing data values. In real-time trading, each stock value is marked in conjunction with a timestamp. In order to extrapolate the correct timestamp from missing incorrect/missing timestamps, every data entry point is checked against the internal system clock to estimate the likely missing timestamp (Narang 2013). However, this timestamp extrapolation method has disadvantages in its high computation cost and slower system response time for huge volumes of data.

There are other ways to handle missing data. Conventional approaches include ignoring missing attributes or fill in missing values with a global constant (Quinlan 1989), with the real possibility of detracting from the quality of pattern(s) discovered based on these values. Another approach was by Grzymala-Busse et al. (2005), that is the closest fit method, where the same attributes from similar cases are used to extrapolate the missing attributes. Other approaches of extrapolation include maximum likelihood, genetic programming, Expectation-Maximization (EM), Expectation-Maximization (EM), and "machine learning approach (such as autoencoder neural network)" (Lakshminarayan et al. 1999).

- **Closest fit Method**

This method determines the closest value of the missing data attribute through the closest fit algorithm based on the same attributes from similar cases. Using the closest fit algorithm, the distance between cases (such as case x and y) are based on the Manhattan distance formula that is given below:

$$\text{distance}(x, y) = \sum_{i=1}^{n} \text{distance}(x_i, y_i)$$

where:

$$\text{distance}(x, y) = \begin{cases} 0 & \text{if } x = y \\ 1 & \text{if } x \text{ and } y \text{ are symbolic and } x \neq y, \text{ or } x = ? \text{ or } y = ? \\ \frac{|x-y|}{r} & \text{if } x \text{ and } y \text{ are numbers and } x \neq y \end{cases}$$

where r denotes the differences between the maximum and minimum of the unknown values of missing numeric values (Grzymala-Busse et al. 2005).

- **Maximum likelihood**:

Maximum likelihood is a statistical method to approximate a missing value based on the probability of independent observations. The beginning point for this approximation is the development of a likelihood function that determines the probability of data as a function of data and its missing value. Allison (2012), the estimation commences with the expression of likelihood function to present the probability of data, as a function of data and its missing value. This function's parameters must maximize the likelihood of the observed value as in the following formulation:

$$L(\theta | Y_{\text{observed}}) = \int f\left(Y_{\text{observed}}, Y_{\text{missing}} | \theta\right) \mathrm{d}Y_{\text{missing}}$$

where Y_{observed} denotes the observed data, Y_{missing} is the missing data, and Θ is the parameter of interest to be predicted (Little and Rubin 1987). Subsequently, likelihood function is expressed by:

$$L(\theta) = \prod_{i=1}^{n} f(y_i | \theta)$$

where $f(y|\theta)$ is the probability density function of the observations y whilst θ is the set of parameters that has to be predicted provided n number of independent observation (Allison 2012). The value of θ must be first determined before a maximum likelihood prediction can be calculated which serves to maximize the likelihood function.

Suppose that there are n independent observation on k variables $(y_1, y_2, \ldots, y_k)$ "with no missing data, the likelihood function "is denoted as:

$$L = \prod_{i=1}^{n} f(y_{i1}, y_{i2}, \ldots, y_{ik}; \theta)$$

However, suppose that data is missing for individual observation i for y_1 and y_2. Then, the likelihood of the individual missing data is dependent on the likelihood of observing other remaining variables such as y_3, …, y_k. Assuming that $y1$ and $y2$ are discrete values, then the joint likelihood is the summation of all possible values of the two variables which have the missing values in the dataset. Consequently, the joint likelihood is denoted as:

$$f_i^*(y_{i3}, \ldots, y_{ik}; \theta) = \sum_{y1} \sum_{y2} f_i(y_{i1}, \ldots, y_{ik}; \theta)$$

As the missing variable are continuous, the joint likelihood is the integral of all potential values of the two variable that contain the missing values in the dataset. Thus, the joint likelihood is expressed as:

$$f_i^*(y_{i3}, \ldots, y_{ik}; \theta) = \int_{y1} \int_{y2} f_i(y_{i1}, y_{i2}, \ldots, y_{ik}) dy_2 dy_1$$

Because each observation adds to the determination of the likelihood function, then the summation (integral) is calculated over the missing values in the dataset. The overall probability is denoted as the product of all observations. An example, if there are x observations with complete data and n-x observations with data missing on y_1 and y_2, the probability function for the full dataset is expressed as:

$$L = \prod_{i=1}^{x} f(y_{i1}, y_{i2}, \ldots, y_{ik}; \theta) \prod_{x+1}^{n} f_i^*(y_{i3}, \ldots, y_{ik}; \theta)$$

The advantages of using the maximum likelihood method to extrapolate missing values are that this method produces approximations that are consistent (in that it produces the same or almost the same unbiased results for a selected large dataset); it is asymptotically efficient (in that there is minimal sample inconsistency which denotes a high level of efficiency in the missing value dataset); and it is asymptotically normal (Allison 2012).

In Fig. 1, the maximum likelihood algorithm, with known variance parameter of sigma, is tested using several small but representative sets of missing value matrices with some rows containing no missing values, other containing one missing value, and still others containing several missing values.

5 Conclusion

The chapter introduced the concept of big data with its characteristics namely velocity, volume, and variety. It introduces the phases of big data management, which includes data cleansing and mining. Techniques that are used during some of these phases are presented. A new category of algorithm, bio-inspired algorithms, is introduced with several example algorithms based on the behaviour of different species of animals explained. Following Zang's et al. (2010) rules for the development of a bio-inspired algorithm, a new algorithm, KSA, is shown with its phases of descriptive animal behaviour, mathematical modelling of this behaviour, algorithmic development, and finally testing with results.

In this chapter, we chose a particular step of data cleansing, extrapolating missing values, of the big data management stages to demonstrate how bio-inspired algorithms work.

Key Terminology & Definitions

Big data—is a definition that describes huge volume and complicated data sets from various heterogeneous sources. Big data is often known by its characteristics of velocity, volume, value, veracity, and variety.

Bio-inspired—refers to an approach that mimics the social behaviour of birds/animals. Bio-inspired search algorithms may be characterized by randomization, efficient local searches, and the discovering of the global best possible solution.

Data imputation—is replacing missing data with substituted values.

References

Abdella, M., & Marwala, T. (2006). The use of genetic algorithms and neural networks to approximate missing data in database. *Computing and Informatics, 24,* 1001–1013.

Agbehadji, I. E. (2011). Solution to the travel salesman problem, using omicron genetic algorithm. Case study: Tour of national health insurance schemes in the Brong Ahafo region of Ghana. M. Sc. (Industrial Mathematics) Thesis. Kwame Nkrumah University of Science and Technology. Available https://doi.org/10.13140/rg.2.1.2322.7281.

Agbehadji, I. E., Fong, S., & Millham, R. C. (2016a). Wolf Search Algorithm for Numeric Association Rule Mining.

Agbehadji, I. E., Millham, R., & Fong, S. (2016b). Wolf search algorithm for numeric association rule mining. In *2016 IEEE International Conference on Cloud Computing and Big Data Analysis (ICCCBDA 2016).* Chengdu, China. https://doi.org/10.1109/ICCCBDA.2016.7529549.

Allison, P. D. (2012). *Handling missing data by maximum likelihood. Statistical horizons.* PA, USA: Haverford.

Banupriya, S., & Vijayadeepa, V. (2015). Data flow of motivated data using heterogeneous method for complexity reduction. *International Journal of Innovative Research in Computer and Communication Engineering, 2*(9).

Cha, S. H., & Tappert, C. C. (2009). A genetic algorithm for constructing compact binary decision trees. *Journal of Pattern Recognition Research, 4*(1), 1–13.

Dorigo, M., & Gambardella, L. M. (1997). Ant colony system: A cooperative learning approach to the traveling salesman problem. *IEEE Transactions on Evolutionary Computation, 1*(1), 53–66.

Dorigo, M., Birattari, M., & Stutzle, T. (2006). Ant colony optimization. *IEEE Computational Intelligence Magazine, 1*(4), 28–39.

Fister, I. J., Fister, D., Fong, S., & Yang, X.-S. (2014). Towards the self-adaptation of the bat algorithm. In *Proceedings of the IASTED International Conference Artificial Intelligence and Applications (AIA 2014),* February 17–19, 2014 Innsbruck, Austria.

Fong, S. J. (2016). *Meta-Zoo heuristic algorithms* (p. 2016). Islamabad, Pakistan: INTECH.
Grzymala-Busse, J. W., Goodwing, L. K., & Zheng, X. (2005). Handling missing attribute values in Preterm birth data sets.
Honkavaara, J., Koivula, M., Korpimäki, E., Siitari, H., & Viitala, J. (2002). Ultraviolet vision and foraging in terrestrial vertebrates. *Oikos, 98*(3), 505–511.
Kennedy, J., & Eberhart, R. C. (1995). Particle swarm optimization. In *Proceedings of IEEE International Conference on Neural Networks* (pp. 1942–1948), Piscataway, NJ.
Krause, J., Cordeiro, J., Parpinelli, R. S., & Lopes, H. S. (2013). A survey of swarm algorithms applied to discrete optimization problems. *Swarm intelligence and bio-inspired computation: Theory and applications* (pp. 169–191). Elsevier Science & Technology Books.
Kumar, R. (2015). Grey wolf optimizer (GWO). Available https://drrajeshkumar.files.wordpress.com/2015/05/wolf-algorithm.pdf. Accessed 3 May 2017.
Lakshminarayan, K., Harp, S. A., & Samad, T. (1999). Imputation of missing data in industrial databases. *Applied Intelligence, 11,* 259–275.
Little, R. J. A., & Rubin, D. B. (1987). *Statistical analysis with missing data.* New York: Wiley.
Longbottom, C., & Bamforth, R. (2013). Optimising the data warehouse. *Dealing with large volumes of mixed data to give better business insights.* Quocirca.
Narang, R. K. (2013). *Inside the black box: A simple guide to quantitative and high frequency trading*, 2nd ed. Wiley: USA. Available: https://leseprobe.buch.de/imagesadb/78/04/78041046-b4fd-4cae-b31d-3cb2a2e67301.pdf Accessed 20 May 2018.
Quinlan, J. R. (1989). Unknown attribute values in induction. In *Proceedings of the Sixth International Workshop on Machine Learning* (pp. 164–168). Ithaca, N.Y.: Morgan Kaufmann.
Selvaraj, C., Kumar, R. S., & Karnan, M. (2014). A survey on application of bio-inspired algorithms. *International Journal of Computer Science and Information Technologies, 5*(1), 366–370.
Shrubb, M. (1982). The hunting behaviour of some farmland Kestrels. *Bird Study, 29*(2), 121–128.
Spencer, R. L. (2002). Introduction to matlab. Available https://www.physics.byu.edu/courses/computational/phys330/matlab.pdf Accessed 10 Sept 2017.
Tang, R., Fong, S., Yang, X.-S., & Deb, S. (2012). Wolf search algorithm with ephemeral memory. In *2012 Seventh International Conference on Digital Information Management (ICDIM)* (pp. 165–172), 22–24 August 2012, Macau. https://doi.org/10.1109/icdim.2012.6360147.
Varland, D.E. (1991). Behavior and ecology of post-fledging American Kestrels.
Vlachos, C., Bakaloudis, D., Chatzinikos, E., Papadopoulos, T., & Tsalagas, D. (2003). Aerial hunting behaviour of the lesser kestrel falco naumanni during the breeding season in thessaly (Greece). *Acta Ornithologica, 38*(2), 129–134. Available: http://www.bioone.org/doi/pdf/10.3161/068.038.0210 Accessed 10 Sept 2016.
Yang, X-S. (2010a). Firefly algorithms for multimodal optimization.
Yang, X. S. (2010b). A new metaheuristic bat-inspired algorithm. In *Nature inspired cooperative strategies for optimization (NICSO 2010)* (pp. 65–74).
Yang, X. S. (2010c). Firefly algorithm, stochastic test functions and design optimisation. *International Journal of Bio-Inspired Computation, 2*(2), 78–84.
Yang, X. S., & Deb, S. (2009, December). Cuckoo search via Lévy flights. In *Nature & Biologically Inspired Computing, 2009. NaBIC 2009. World Congress on* (pp. 210–214). IEEE.
Zang, H., Zhang, S., & Hapeshi, K. (2010). A review of nature-inspired algorithms. *Journal of Bionic Engineering, 7,* S232–S237.

Richard Millham is currently an Associate Professor at Durban University of Technology in Durban, South Africa. After thirteen years of industrial experience, he switched to academe and has worked at universities in Ghana, South Sudan, Scotland, and Bahamas. His research interests include software evolution, aspects of cloud computing with m-interaction, big data, data streaming, fog and edge analytics, and aspects of the Internet of Things. He is a Chartered Engineer (UK), a Chartered Engineer Assessor and Senior Member of IEEE.

Israel Edem Agbehadji graduated from the Catholic University college of Ghana with B.Sc. Computer Science in 2007, M.Sc. Industrial Mathematics from the Kwame Nkrumah University of Science and Technology in 2011 and Ph.D. Information Technology from Durban University of Technology (DUT), South Africa, in 2019. He is a member of ICT Society of DUT Research group in the Faculty of Accounting and Informatics; and *IEEE* member. He lectured undergraduate courses in both DUT, South Africa, and a private university, Ghana. Also, he supervised several undergraduate research projects. Prior to his academic career, he took up various managerial positions as the management information systems manager for National Health Insurance Scheme; the postgraduate degree programme manager in a private university in Ghana. Currently, he works as a Postdoctoral Research Fellow, DUT-South Africa, on joint collaboration research project between South Africa and South Korea. His research interests include big data analytics, Internet of Things (IoT), fog computing and optimization algorithms.

Hongji Yang graduated with a Ph.D. in Software Engineering from Durham University, England with his M.Sc. and B.Sc. in Computer Science completed at Jilin University in China. With over 400 publications, he is full professor at the University of Leicester in England. Prof Yang has been an IEEE Computer Society Golden Core Member since 2010, an EPSRC peer review college member since 2003, and Editor in Chief of the International Journal of Creative Computing.

Chapter 2
Parameter Tuning onto Recurrent Neural Network and Long Short-Term Memory (RNN-LSTM) Network for Feature Selection in Classification of High-Dimensional Bioinformatics Datasets

Richard Millham, Israel Edem Agbehadji, and Hongji Yang

1 Introduction

The introduction describes the characteristics of big data, review on method and search strategies for feature selection. With the current dispensation of big data, reducing the volumes of dataset may be "achieved by selecting relevant features for classification. Moreover, big data is also characterized by velocity, value, veracity and variety. The characteristic of velocity relates to "how fast incoming data need to be processed and how quickly the receiver of information needs the results from the processing system" (Longbottom and Bamforth 2013); the characteristic of volume refers to the amount of data for processing; the characteristic of value refers to what a user will gain from data analysis. Other characteristics of big data include "variety and veracity." The characteristic of variety looks at "different structures of data such as text and images, while the characteristic of veracity focuses on authenticity of the data source." While these characteristics (i.e., volume, value, variety and veracity) are significant in any big data analytics, it is important to reduce the volume of dataset and produce value (relevant and useful features) with reduced computational cost given

R. Millham (✉) · I. E. Agbehadji
ICT and Society Research Group, Durban University of Technology, Durban, South Africa
e-mail: richardm1@dut.ac.za

I. E. Agbehadji
e-mail: israeldel2006@gmail.com

H. Yang
Department of Informatics, University of Leicester, Leicester, England, UK
e-mail: Hongji.Yang@Leicester.ac.uk

S. J. Fong and R. C. Millham (eds.), *Bio-inspired Algorithms for Data Streaming and Visualization, Big Data Management, and Fog Computing*, Springer Tracts in Nature-Inspired Computing,
https://doi.org/10.1007/978-981-15-6695-0_2

the rapid nature at which data is generated from the big data environment. Hence, a new and efficient algorithm is required to manage the volume, handle velocity and produce value from data.

An aspect of velocity characteristics is the use of parameter to speed up training of network. Weights parameter setting is recognized as an effective approach as it "influence not only speed of convergence but also the probability of convergence." Thus, using "too small or too large values could speed the learning, but at the same time, it may end up performing worse. In addition, the number of iterations of the training algorithm and the convergence time would vary depending on the initialized value" of parameters.

In this chapter, we propose a search strategy to address these issues of volume, velocity and value, by exploring the behavior of kestrel bird in performing random encircling and imitation in finding weight parameter for deep learning.

2 Feature Selection

Feature selection helps to select relevant features from large number of features and ignore irrelevant features with little value on output feature set. Generally, features are characterized as relevant, irrelevant and redundant. A feature is said to be a relevant feature when it has an influence on output features and its role cannot be assumed by other features. Irrelevant feature is a feature that influences an outcome of a result. On the other hand, redundant feature is a feature that takes the role of another feature in subset. Binh and Bing (2014) indicated that in the feature selection process, the performance of search algorithm is more significant than the number of feature that are selected, as this can be a attributed to the fact that search algorithms should use less time to select an approximate feature than spend an extensive amount of time to select some number of features which then could lose its usefulness. This suggests that time used by search strategies is very fundamental in the process of feature subset generation (Waad et al. 2013). There are different techniques to employ in developing search strategies which can be categorized as the filter method (Dash and Liu 1997), wrapper method (Dash and Liu 1997) and embedded method (Kumar and Minz 2014).

2.1 Filter Method

The first category which is the filter method finds the relevance of a feature (Dash and Liu 1997) in a class by evaluating a feature without a learning algorithm. Classification algorithms that adopt the filter method "evaluate the goodness of a feature" and rank features based on distance measure (Qui 2017), information measure (Elisseeff and Guyon 2003) and dependency measure (Almuallim and Dietterich 1994). The distance measure finds the difference in values between two features and if the difference is zero then the features are "indistinguishable." The information measure finds the "information gain from a feature" as a function of the difference between uncertainties that is prior and posterior value of information that is gained. Consequently, a feature is selected if the "information gain from one feature is greater than the other feature" (Ben-Bassat 1982). The dependency measure, also referred to as the correlation measure or similarity measure, predicts the value of one feature from the value of another feature. A feature is predicted based on how strong it is associated to a class of features.

3 Wrapper Method

The second category is the wrapper method (Hall 2000) which uses a learning algorithm to learn from every possible feature subset, trains the selected subset and evaluates its usefulness (Dash and Liu 1997; Liu and Yu 2005). The selected features are ranked according to the usefulness and predictive ability of the classification algorithm that is measured in terms of performance (Kohavi and John 1996). The performance measures uses "statistical re-sampling technique called cross validation (CV)" which measures classification accuracy of results. Although accuracy of results is guaranteed, high computational time is required for learning and training (Uncu and Turksen 2007) when big datasets are involved. Some search techniques used in wrapper method are sequential search, exhaustive search and random search (Dash and Liu 1997). The sequential search strategy uses forward selection and backward elimination to iteratively add or remove features. Initially, the forward selection algorithm starts an iteration with an empty dataset. At each iteration, best features are sequentially selected by an objective function and added to the empty initial dataset until there is no more features to be selected (Whitney 1971). When the search is being performed, a counter is set to count the number of updates that happens on a subset. The challenge of this search algorithm is that once features are selected into a subset, when the feature becomes obsolete or not useful, it cannot be removed from the subset and this could lead to loss of optimal subsets (Ben-Bassat 1982) even if the search gives solution is a reasonable amount of time. On the other hand, the backward selection algorithm starts with a full dataset of features, and during iteration, objective function is applied to perform sequential backward update of the dataset by removing the least significant features that do not met a set criteria (Marill 1963) from a subset. When the algorithm removes least significant feature, counter is used to count the number of updates that were perform on the subset. The advantage of backward selection algorithm is that it guarantees a quick convergence to optimal solution. Whilst, an exhaustive search performs a complete search of the entire feature subset and then selects the possible optimal results (Waad et al. 2013). When the number of features grow exponentially, the search takes more computational time (Aboudi and Benhlima 2016), thus leading to low performance results. The random search strategies perform a search by randomly finding subsets of features (Dash and Liu 1997). The advantage of random search strategy over sequential and exhaustive search is the reduction in computation cost. The random search strategy (also referred to as population based search) is a meta-heuristic optimization approach which is based on the principle of evolution in searching for a better solution in a population. These evolutionary principles are best known for their global search abilities. The search process starts within random initialization of a solution or candidate solution, iteratively updates the population that satisfies a fitness function and terminates when a stopping criteria is met.

3.1 Embedded Method

The embedded method that selects feature by putting data into two sets: training and validation sets. When variables that define features are selected for training, the need to retrain a variable that can be used to predict every variable subset feature is avoided (Kumar and Minz 2014) and this makes the embedded method able to reach a fast solution. However, predictor variable selection is model specific meaning with each feature selection model being used, different variables have to be defined, thus making the embedded method a model specific.

4 Machine Learning Methods

As mentioned earlier, among the traditional approach to learning methods/machine learning methods includes artificial neural network (ANN) and support vector machine (SVM).

4.1 Artificial Neural Network (ANN)

The artificial neural network is interlink of "group of nodes (neurons)" where each "node receives inputs from other nodes and assigns weights between nodes to adapt so that the whole network learns to perform useful computations" (Bishop 2006). Mostly, algorithms based on ANN are slow learners in that they require many iterations over the training set before choosing its parameter (Aamodt 2015) leading to high computation. The neural network structure and learning algorithms uses perceptron neural network (i.e., an algorithm for supervised classification) and back-propagation. The advantage of a learning algorithm is that it helps in adapting weights of a neural network by minimizing error between a desired output and an actual output. The aim of back-propagation "algorithm is to train multilayer neural networks by computing error derivatives in hidden activities in hidden layers and updating weights accordingly" (Kim 2013). The back-propagation algorithm uses "gradient descent to adjust the connections between units within the layers such that any given input tends to produce a corresponding output" (Marcus 2018).

4.2 Support Vector Machine (SVM)

The support vector machine (SVM) performs "classification by constructing an n-dimensional hyper-plane that optimally separates data into two categories" (Boser et al. 1992). In the processing of constructing the hyper-plane, the SVM creates a validation set that determines the value of parameter for the training algorithm to find the maximum margin separating feature space in hyper-plane between two classes of points. Although this separation into two classes may look quite simple and easy (i.e., involving no local minima), it requires the use of "good" function approximator (i.e., kernel function) in finding a parameter when large volumes of data are used in training and this results in high computational cost (Lin 2006).

The challenge with the traditional approach to learning (such as ANN and SVM) led to the "concept of deep learning which historically originated from artificial neural network" (Deng and Yu 2013).

5 Deep Learning

Deep learning is an "aspect of machine learning where learning is done in hierarchy." In this context, "higher-level features can be defined from lower-level features and vice versa" (Deng and Yu 2013; Li 2013). The hierarchical representation of deep learning structure enables classification using multiple layers. Hence, deep learning relates some structure in neural network (Marcus 2018).

The neural networks used in deep learning consist of a set of input units (examples are pixels or words), "multiple hidden layers (the more such layers, the deeper a network is said to be) containing hidden units (also known as nodes or neurons), and a set output units, with connections running between those nodes" (Marcus 2018) to form a map between inputs and outputs. The map shows the complex representation of large data and provides an efficient way to optimize a complex system such that the test dataset closely resembles the training set. This close resemblance suggests a minimization of deviations between test and training set in large dataset; therefore, deep learning is a way to optimize complex systems to map inputs and outputs, given a sufficient amount of data (Marcus 2018).

In principle, deep learning uses multiple hidden layers which are nonlinear, and mostly different parameters are employed to learn from hidden layers (Patel et al. 2015). The categories of deep learning methods for classification are discriminative models/supervised-learning (e.g., deep neural networks (DNN), recurrent neural networks (RNN), convolutional neural networks (CNN) etc.); "generative/unsupervised models" (e.g., restricted Boltzmann machine (RBM), deep belief networks (DBN), deep Boltzmann machines (DBM), regularized autoencoders, etc.).

5.1 Deep Neural Network (DNN)

Deep neural network (DNN) is a "multilayer network that has many hidden layers in its structural representation and its weights are fully connected with the model" (Deng 2012). In some instances, recurrent neural networks (RNN) which is a discriminative model is used as generative model, thus enables the output results to be used as input data in a model (Deng 2012). Recurrent nets (RNNs) have been applied on "sequential data such as text and speech" (LeCun et al. 2015) to scale up large text and speech recognition. Although learning of parameters in RNN has been improved through the use of information flow in bi-directional RNN and a cell of long short-term memory (LSTM) (Deng and Yu 2013), the challenge is that the back-propagated gradients either "grow or shrink (i.e., decay exponentially in the number of layers) at each time step" (Tian and Fong 2016), so over many time steps it typically explodes or vanishes (i.e., increase out of bound or decrease at each iteration) (LeCun et al. 2015). Several methods to solve the exploding and shrinking of a learned parameter include primal-dual training method, cross entropy (Deng and Chen 2014), echo state network, sigmoid as activation functions (Sohangir et al. 2018), etc. While the primal-dual training method was formulated as an optimization problem, "the cross entropy is maximized, subject to the condition that the infinity norm of the recurrent matrix of the RNN is less than a fixed value to guarantee the stability of RNN dynamics" (Deng and Yu 2013). In the echo state network, the "output layers are fixed to be linear instead of nonlinear" and "where the recurrent matrices are designed but not learned." Similarly, the "input matrices are also fixed and not learned, due partly to the difficulty of learning." While sigmoid functions are mathematical expression that defines output of a neural network given a set of data inputs. Meanwhile, the use of LSTM enables networks to remember inputs for a long time using a memory cell (LeCun et al. 2015). LSTM networks have subsequently proved to be more effective especially when they have several layers for each time step (LeCun et al. 2015).

5.2 Convolutional Neural Network (CNN)

Convolutional neural network (CNN) shares many weights, and pooling outputs from different layers, thereby reducing the data rate from the lower layers of the network. The CNN has been found highly effective in computer vision, image recognition (LeCun et al. 1998; Krizhevsky et al. 2012) and speech recognition (Deng and Yu 2013; Abdel-Hamid et al. 2012; Abdel-Hamid and Deng 2013; Sainath et al. 2013) where it can analyze internal structures of complex data through convoluted layers.

Similarly, CNN has been also found highly effective on text data such as sentence modeling, search engines, in systems for tagging (Weston et al. 2014; Collobert et al. 2011), sentiment analysis (Sohangir et al. 2018) and stock market price prediction (Aamodt 2015). "Convolutional deep belief networks help to scale up to high-dimensional dataset" (Lee et al. 2009). By applying this network to images, it shows good performance in several visual recognition tasks (Lee et al. 2009). "Deep convolutional nets have brought about breakthroughs in processing images, video, speech and audio."

5.3 Restricted Boltzmann Machine (RBM)

Restricted Boltzmann machine (RBM) is often "considered as a special Boltzmann machine, which has both visible units and hidden units as layers, with no visible–visible or hidden–hidden connections" (Deng 2011). The deep Boltzmann machine (DBM) has hidden units organized in a deep layered manner, where only adjacent layers are connected, and there are no visible–visible or hidden–hidden connections within the same layer. The deep belief network (DBN) is "probabilistic generative models that composed of multiple layers of stochastic, hidden variables." The DBN has top two layers in its structure that are undirected with symmetric connections between them. The DBN also has a lower layer that is directed with connections from layers above it. Another generative model is the deep auto-encoder which is a DNN whose output target is the data input itself, often pre-trained with DBN or using "distorted training data to regularize the learning." Table 1 shows a summary on related work on deep learning as follows:

It is observed that current research has applied deep learning to different search domains such as image processing, stock trading, character recognition in sequential text analysis, etc. This shows the capabilities of the deep learning methods.

The difference between supervised and unsupervised learning models that were discussed earlier is that, in a supervised learning, a pre-classified example of features is "available for learning and the task is to build a (classification or prediction) model that will work on unseen examples; whereas in an unsupervised learning," there is neither pre-classified example nor feedback (this technique is suitable for clustering and segmentation tasks) to the learning model (Berka and Rauch 2010). In training these networks, a gradient descent algorithm is employed, which allows the back-propagation algorithm to compute a vector representation using an objective function (Le 2015). However, back-propagation alone is not efficient because of its being stacked in a "local optima in the non-convex objective function" (Patel et al. 2015). In order to avoid this local optimum, meta-heuristic search methods were adopted in building classifiers when search space is growing exponentially. The advantage is that it enhances computational efficiency and quality of selecting useful and relevant features (Li et al. 2017). Meta-heuristic algorithms that have been integrated with traditional machine learning methods include the following as indicated by Fong et al. (2013), Zar (1999) in Table 2.

Table 1 Deep learning methods and problem domain

Deep learning method	Search/problem domain	Author(s)
Convolutional deep belief networks	Unsupervised feature learning for audio classification	Honglak Lee, Yan Largman, Peter Pham and Andrew Y. Ng.
Convolutional deep belief networks	Scalable unsupervised learning of hierarchical representations	Lee et al. (2009)
Deep convolutional neural networks (DCNN)	Huge number of high resolution images	Krizhevsky et al. (2012)
Deep neural network	The classification of stock and prediction of prices.	Batres-Estrada (2015)
Deep neural network-hidden markov models (DNN-HMMs).	Discovering features in speech signals	Graves and Jaitly (2014)
Train the CNN architecture based on the back-propagation algorithm	Character recognition in sequential text	31
Deep convolutional neural network	Event-driven stock prediction	Ding et al. (2015)
Convolutional neural network	Stock trading	Siripurapu (2015)

Table 2 Meta-heuristic algorithms integrated with traditional method

Authors	Traditional methods of classification	Meta-heuristic/bio-inspired algorithm	Search domain
Ferchichi (2009)	Support vector machine	Tabu search, genetic algorithm	Urban transport
Alper (2010)	Logistic regression	Particle swarm optimization, scatter search, Tabu search	General
Unler et al. (2010)	Support vector machine	Particle swarm optimization	General
Abd-Alsabour (2012)	Support vector machine	ACO	General
Liang et al. (2012)	Rough set	Scatter search	Credit scoring
Al-Ani (2006)	Artificial neural network	ACO	Speech, image
Fong et al. (2013)	Neural network	Wolf search algorithm	General

It is observed from Table 2 that research is focused on traditional machine learning methods with meta-heuristic search methods. However, with the current dispensation of very large volumes data, traditional machine learning methods are not suitable because of the risk of being stuck in local optima and chances are that same results might be recorded as more data is generated which might not give an accurate result on feature selection for a classification problem.

6 Meta-Heuristic/Bio-Inspired Algorithms

Among the population-based/random search algorithms for feature selection in classification problems are genetic algorithm (GA) (Agbehadji 2011), ant colony optimization (ACO) (Cambardella 1997), particle swarm optimization (PSO) (Kennedy and Eberhart 1995) and wolf search algorithm (WSA) (Tang et al. 2012).

6.1 Genetic Algorithms

Genetic algorithm is an evolutionary method which depends on "natural selection" (Darwin, 1868 as cited by Agbehadji 2011). The stronger the genetic composition of an individual, the more its capable to withstand competitions in its environment. The search process makes the genetic algorithm to be adapt to any given search space (Holland 1975), as cited by Agbehadji (2011). This search process uses what is called operators such as crossover, mutation and selection methods to search for a global optimal results/solution that meets a fitness value. During the search, there is an initial guess which is improved through "evolution by comparing the fitness of the initial generation of population with the fitness obtained after" application of "operators to the current population until the final optimal value is produced" (Agbehadji 2011).

6.2 Ant Colony Optimization (ACO)

The ant colony optimization (ACO) (Cambardella 1997) mimics the foraging capabilities of ants when searching food in its natural environment. When ants search for food the deposit a substance called pheromone to assist other ants to locate the path to a food source. The quantity of pheromone is based on distance, quantity and quality of food source (Al-Ani 2007). The challenge of pheromone substance is that it does not last longer (Stützle and Dorigo 2002). Thus, ants make probabilistic decisions which enables its to update their pheromone trail (Al-Ani 2007) so as to explore larger search space.

6.3 Wolf Search Algorithm (WSA)

Wolf search algorithm (WSA) is based on the preying behavior of wolf (Tang et al. 2012). The wolf is able to use scent marks to demarcate its territory and communicate with other wolves of the pack (Agbehadji et al. 2016).

6.4 Particle Swarm

Particle swarm is a bio-inspired method based on the swarm behavior such as fish and bird schooling in nature (Kennedy and Eberhart 1995). The swarm behavior is expressed in terms of how particles adapt, exchange information and make decision on change of velocity and position within a space based on position of other neighboring particles. The search characteristics of particle swarm involve initialization of particles and several iterations are performed to update position of each particle depending on the value assigned to its velocity and combined to its best previous own position and the position of the best element among the global population of particles (Aboudi and Benhlima 2016). The advantage of particle swarm's behavior is the ability for local interaction among particles that leads to an emergent behavior, which relates to global behavior of particles in a population (Krause et al. 2013). Particle swarm methods are computationally less expensive which makes it more attractive and effective for feature selection. Again, each particle discovers the best feature combination as they move in a population. When applying particle swarm to any feature selection problem, it is important to define a threshold value during initialization to decide which feature is selected or discarded. Often, it is difficult for a user to explicitly set a threshold since it might influence performance of the algorithm (Aboudi and Benhlima 2016). The initialization strategy proposed by (Xue et al. 2014) adopts sequential selection algorithm to guarantee accuracy of classification and to show the number of features selected (Aboudi and Benhlima 2016). The novelty of this chapter is the combination of deep learning search method with the proposed bio-inspired/meta-heuristic/population-based search algorithm to avoid possibility of being stuck in local optima in the large volumes of dataset for feature selection. In this chapter, we will propose a search strategy to avoid being trapped in local optima when the search space grow exponentially for each time step (iteration) by exploring the behavior of kestrel bird in performing random encircling and imitation in finding weight parameter for deep learning.

7 Proposed Bio-Inspired Method: Kestrel-Based Algorithm with Imitation

The chapter one of this book considered the mathematical formulation and algorithm of kestrel bird. This section models the imitative behavior of kestrel bird. Basically, kestrel birds are territorial and hunt individually rather than hunt collectively (Shrubb 1982; Varland 1991; Vlachos et al. 2003; Honkavaara et al. 2002; Kumar 2015; Spencer 2002). As a consequence, a model by that depicts the collective behavior of birds for feature similarity selection could not be applied (Cui et al. 2006). Since kestrels are imitative, it implies that a well-adapted kestrel would perform action appropriate to its environment, while other kestrels that are not well-adapted imitate and remember the successful actions. The imitation behavior reduces learning and improves upon the skills of less adapted kestrels. A kestrel that is not well adapted to an environment imitates the behavior of well-adapted kestrels.

A kestrel is most likely to take a random step that better imitates a successful action. The imitation learning is an approach to skill acquisition (Englert et al. 2013) where a function is expressed to transfer skills to lesser-adapted kestrels. The imitation learning rate determines how much to update the weights parameter during the training (Kim 2013). Having a large value for the learning rate makes the lesser-adapted kestrels to quickly learn, but it may not converge or result in poor performance. On the other hand, if the value of learning rate is too small, it is inefficient as it takes too much time to train lesser-adapted kestrels. In our approach, we imitated the position at which a kestrel can copy an action from a distance. Hence, a short distance enables a high imitation. The imitation is mathematically expressed and applied to select similar features into a subset. A similarity value $\mathrm{Sim}_{\mathrm{value}(O,T)}$ that helps with the selection of similar features is expressed by:

$$\mathrm{Sim}_{\mathrm{value}(O,T)} = e^{\left(-\frac{\sum |O_i - E_i|^2}{n}\right)} \tag{1}$$

where "n is the total number of features, $|(O_i - E_i)|$ represents the deviation between two features where O is the observed, E_i is an estimate that is the velocity of kestrel. Since the deviation is calculated for each feature dimension and there is the possibility of large volume of features in dataset, each time a deviation is calculated only the minimum is selected (the rest of the dimension is discarded), thus, enabling it to allow the handling of different problem to different scale of dimension of data" (Blum and Langley 1997). In cases where features imitated are not similar (i.e., dissimilarity), it is expressed by:

$$\mathrm{dis\_sim}_{\mathrm{value}\,(O,T)} = 1 - \mathrm{Sim}_{\mathrm{value}\,(O,T)} \tag{2}$$

The fitness function, which is similar to fitness function formulation used by (Mafarja and Mirjalili 2018), helps to evaluate intermediate results is expressed in terms of classification error based on RNN with LSTM. The fitness function is formulated as:

$$\mathrm{fitness} = \rho * \mathrm{Sim}_{\mathrm{value}\,(O,T)} + \mathrm{dis\_sim}_{\mathrm{value}\,(O,T)} * \rho \tag{3}$$

where $\rho \in (0, 1)$ is a parameter that controls the chances of imitating features that are dissimilar, C_{error} is the classification error of a RNN with LSTM classifier and $\mathrm{Sim}_{\mathrm{value}\,(O,T)}$ refers to feature similarity value.

The RNN with LSTM is used to make decision on classification accuracy so as to scale up large data. In order to select the best subset of feature, the study applied this concept, which states that the "less the number of features in a subset and the higher the classification accuracy, the better the solution" (Mafarja and Mirjalili 2018). The proposed kestrel-based search algorithm with imitation for feature selection is expressed in Table 3 as follows:

Table 3 Proposed algorithm

Set parameters
Initialize population of n Kestrels using equation.
Start iteration (loop until termination criterion is met)
    Generate new population using random encircling
    Compute the velocity of each kestrel using
    Evaluate fitness of each solution (equation 3)
    Update encircling position for each Kestrel for all i=1 to n
End loop
Output optimal results

The formulation on kestrel algorithm also adopts aspect of swarm behavior in terms of "individual searching, moving to better position, and fitness evaluation" (Agbehadji et al. 2016). However, "what makes kestrel distinctive is the individual hunt through its random encircling of prey and its imitation of the best individual kestrel. Since kestrel hunts individually and imitates the best features of successful individual kestrel, it suggests that kestrels are able to remember the best solution from a particular search space and continue to improve upon initial solution until the final best is reached. In kestrel search algorithm, each search agent checks the brightness of trail substances using the half-life period; random encircling of each position of a prey before moving with a velocity; imitates the velocity of other kestrels so that each kestrel will swarm to the best skilled kestrel" (Agbehadji et al. 2016). The advantage of KSA is that it adapts to changes in its environment (such as change in distance), thus making it applicable to dynamic and changing data environment.

In comparison of the unique characteristics of the kestrel algorithm (Agbehadji et al. 2016) with the PSO and ACO algorithms, the following can be stated when performing local search and update of current best solution: In PSO, the swarming particles have velocities. So, in PSO, we need not only to update their positions, but also their velocities. Recording the best local solution and global solution in each generation is required. In ACO, each search agent updates their pheromone substances, rate of evaporation and the cost function in order to move into search for food. In KSA, each search agent applies random encircling, imitates the best position of other search agents in each iteration and the half-life of each trail tells kestrel how long its best position can last. This informs kestrels the next position to use for the next random encircling.

8 Experimental Setup

The proposed algorithmic structure was implemented in MATLAB 2018A. For the purpose of ensuring that best solution (in terms of optimized parameters) is selected as learning parameter for training the RNN with LSTM network classifier (with

100 hidden layers), 100 iterations were performed. Similarly, 100 epochs were performed in the LSTM network as suggested by (Batres-Estrada 2015) that it guarantees optimum results on classification accuracy. The authors of (Batres-Estrada 2015) indicated that choosing a small value as learning rate makes the interactions in weight space smooth, but at the cost of longer learning rate. Similarly, choosing a large learning rate parameter makes the adjustment too large which makes the network unstable (i.e., the deep learning network) in terms of performance. In this experiment, the stability of the network is maintained by allowing the neurons in the input and out layer to learn at the same rate, smaller learning rate (Batres-Estrada 2015). The use of smaller or optimized learning rate/parameter was achieved by the use of meta-heuristic algorithms such as KSA. The optimized results from the meta-heuristic algorithms and the respective results on classification accuracy are the criteria to evaluate each meta-heuristic algorithm used in the experiment for classification of features. The solution from each meta-heuristic algorithm is considered as best solution if it has higher classification accuracy. The initial parameters for each meta-heuristic algorithm are defined as suggested by authors of the algorithms as best parameters that guarantee an optimal solution (Table 4).

The meta-heuristic algorithms namely PSO, ACO, WSA-MP and BAT which were discussed in literature are used to benchmark the performance of KSA and the best algorithm is selected based on the accuracy of the classification results. During the experiment, nine standard benchmark dataset (i.e., Arizona State University's biological dataset) was used. These datasets were chosen because it represents a standard benchmark dataset with continuous data for experimental research that are suitable for this research work. These parameters were tested on the benchmark datasets shown on Table 5.

Table 4 Algorithm and initial parameters

Algorithm	Initial parameter
KSA	fb = 0.97; % frequency of bobbing z_{min} = 0.2; % perched parameter z_{max} = 0.8; % flight parameter Half-life = 0.5; % half-life parameter Dissimilarity = 0.2% dissimilarity parameter Similarity = 0.8% similarity parameter
PSO	$w = 1$; %inertia weight $c1 = 2.5$; %personal/cognitive learning coefficient $c2 = 2.0$; %global/social learning coefficient
ACO	$\alpha = 1$;%pheromone exponential weight $\rho = 0.05$;%evaporation rate
BAT	$\beta = 1$; % random vector which is drawn from a uniform distribution [0, 1] $A = 1$; %loudness (constant or decreasing) $r = 1$; %pulse rate (constant or decreasing)
WSA-MP	$v = 1$; % radius of the visual range pa = 0.25; %escape possibility; how frequently an enemy appears $\alpha = 0.2$; % velocity factor (α) of wolf

Table 5 Benchmark datasets and number of features in dataset

	Dataset	#of Instances	#of classes	#of features in original dataset
1	Allaml	72	2	7129
2	Carcinom	174	11	9182
3	Gli_85	85	2	22,283
4	Glioma	50	4	4434
5	Lung	203	5	3312
6	Prostate-GE	102	2	5966
7	SMK_CAN_187	187	2	19,993
8	Tox_171	171	4	5748
9	CLL_SUB_111	111	3	11,340

8.1 Experimental Results and Discussion

The minimum learning parameter from the original dataset and classification accuracy helped to evaluate and compare the different meta-heuristic algorithms. 100 iterations were performed by each algorithm to refine parameters for the LSTM network classifier on each dataset (i.e., Arizona State University's biological dataset). Similarly, 100 epochs were performed in the LSTM network as suggested by (Batres-Estrada 2015) that it guarantees optimum results on classification accuracy. Table 6 shows the learning parameter in terms of optimum value of each meta-heuristic algorithm.

Table 6 shows the optimum/minimum learning parameter obtained for each algorithm. It is observed that out of the nine datasets that were used KSA has the best learning parameter in 5 dataset. The best learning parameter for each meta-heuristic algorithm is highlighted in bold. The different learning parameters were fed into LSTM network to determine the performance in terms of classification accuracy by

Table 6 Optimum learning parameter of algorithms

Learning parameter	KSA	BAT	WSA-MP	ACO	PSO
Allaml	**4.0051e−07**	1.232e−07	1.7515e−07	3.3918e−07	1.9675e−06
Carcinom	**1.3557e−07**	1.0401e−07	3.0819e−05	8.7926e−04	0.5123
Gli_85	4.1011	0.032475	3.6925	**0.0053886**	2.2259
Glioma	**2.3177e−06**	3.0567e−05	1.9852e−05	9.9204e−04	0.3797
Lung	**5.1417e−06**	4.4197e−05	3.0857e−05	6.231e−04	0.3373
Prostate-GE	**1.6233e−07**	4.5504e−06	1.0398e−06	3.4663e−05	0.1178
SMK_CAN_187	0.015064	1.338e−05	**4.7188e−05**	2.7294e−05	2.5311
Tox_171	0.16712	**0.0002043**	0.086214	0.0023152	2.2443
CLL_SUB_111	0.82116	0.075597	0.76001	**0.011556**	9.6956

Table 7 Best results on accuracy of classification for each algorithm

Classification accuracy	KSA	BAT	WSA-MP	ACO	PSO
Allaml	0.5633	0.6060	**0.6130**	0.5847	0.4459
Carcinom	**0.7847**	0.7806	0.6908	0.7721	0.7282
Gli_85	0.2000	**0.4353**	0.2004	0.4231	0.3335
Glioma	0.7416	0.7548	0.5063	0.7484	**0.7941**
Lung	0.5754	0.5754	0.5754	0.5754	**0.7318**
Prostate-GE	0.6852	0.6718	0.6147	0.5444	**0.7223**
SMK_CAN_187	**0.6828**	0.6759	0.6585	0.6111	0.2090
Tox_171	**0.7945**	0.6925	0.7880	0.5889	0.2127
CLL_SUB_111	**0.7811**	0.4553	0.7664	0.4259	0.2000
Average	**0.6454**	**0.6275**	**0.6015**	**0.586**	**0.4864**

each algorithm (i.e., a way of knowing which algorithm outperform each other) and the results are shown on Table 7 as follows:

Table 7 shows the classification accuracy using the full dataset and the learning parameter from each algorithm. The classification accuracy for **Allaml** dataset using KSA is 0.56 while WSA-MP is 0.6130. It is observed that the algorithm with best parameter is not the best choice on some dataset. For instance, KSA has the best parameter of **1.6233e−07 on** Prostate-GE dataset but produced a classification accuracy of 0.6852 while BAT has a worst parameter of 0.1178 but produced a classification accuracy of **0.7223.** Hence, a minimum learning parameter does not always guarantee classification accuracy as more features from dataset were imitated. It could be observed that KSA has the highest classification accuracy on four out of nine datasets. This indicates that the proposed algorithm explores and exploits search space efficiently, so as to find best results that produce higher classification accuracy. In order to select features (Mafarja and Mirjalili 2018), indicated that the higher the classification accuracy, the better the solution and hence the less the number of features in a subset. Table 8 shows the dimensions of feature selected by each algorithm.

Table 8 shows the features that were from the respective dataset by each algorithm. It is observed that KSA selected less number of features from **four** datasets, namely Carcinom, SMK_CAN_187, Tox_171 and CLL_SUB_111; PSO selected less feature from **three** datasets, namely Glioma, Lung and Prostate-GE; BAT and WSA-MP selected less number of feature from Gli_85 and Allaml datasets, respectively. This demonstrates that KSA can explore and exploit a search space efficiently and select features that are representative of a dataset.

In this chapter, we conducted a statistical test on classification accuracy of each algorithm to identify the best algorithm. In order not to prejudice which algorithm outperformed each other, the mean of all the algorithms was considered as equal for the statistical analysis.

Table 8 Dimensions of feature selected by each algorithm

Feature selected	KSA	BAT	WSA-MP	ACO	PSO
Allaml	3113	2809	2759	2961	3950
Carcinom	1977	2015	2839	2093	2496
Gli_85	17,826	12,583	17,817	12,855	14,852
Glioma	1146	1087	2189	1116	913
Lung	1406	1406	1406	1406	888
Prostate-GE	1878	1958	2299	2718	1657
SMK_CAN_187	6342	6480	6828	7775	15,814
Tox_171	1181	1768	1219	2363	4525
CLL_SUB_111	2482	6177	2649	6510	9072

8.2 *Statistical Analysis of Experimental Results*

The statistical analysis helped to determine the significance results on classification accuracy from each bio-inspired algorithm (KSA, BAT, WSA-MP, ACO and PSO). In this chapter, we conducted a non-parametric statistical test to assess which of the algorithms have better performance in terms of the classification accuracy. The authors of (García et al. 2007) indicated that non-parametric or distribution-free statistical procedures help to perform "pairwise comparison on related." In a multiple comparison situations such as in this article, the Wilcoxon signed-rank test was applied to test how significant algorithms outperform with respect of detecting the differences in the mean (García et al. 2007) and to find the probability of an error in determining that the median of two comparing algorithms is the same, this probability is referred to as *p*-value (Zar 1999). In applying Wilcoxon test, there is no need to make underlying assumption on the population being used since Wilcoxon test can guarantee to "about 95% (i.e., 0.05 level of significance) of efficiency if the population is normally distributed." The steps in computing Wilcoxon signed-rank test are followed as:

Step 1: "Compute the difference D of paired samples in each algorithm. Any pairs with a difference of 0 are discarded"

Step 2: Find the absolute D.

Step 3: "Compute the rank of signs (R_+ difference and R_- difference) from lowest to highest."

Where sum of ranks is expressed by:

$$\sum R_+ + R_- = \frac{n(n+1)}{2} \tag{21}$$

where n is sample size.

Step 4: "Compute the test statistic T. Thus, $T = \min\{R+, |R-|\}$. Thus, the test statistic T is the smallest value."

Table 9 Test statistics

Comparative algorithms	Z	Asymp. sig. (2-tailed)
BAT—KSA	−0.420	0.674
WSA-MP—KSA	−1.680	0.093
ACO—KSA	−0.980	0.327
PSO—KSA	−1.007	0.314

Step 5: Find the "critical values based on the sample size n." If the T is "less or equal to the critical value at a level of significance (i.e., $\alpha = 0.05$), then a decision is made that algorithms are significantly different" (García et al. 2007). In order to accomplish this, the "Wilcoxon signed-rank table is consulted, using the critical value ($\alpha = 0.05$) and sample size n as parameters," to obtain the value within the table. "If this value is less than the calculated value of the algorithmic comparison, this means that the algorithmic difference is significant."

In order to apply the Wilcoxon signed-rank test, an analysis was performed on classification accuracy and the results are displayed in Table 9 as follows:

Based on the results on test statistics ($p < 0.05$), it shows that the differences between the medians are not statistically significantly different in all the comparative algorithms. For instance, there is no statistically significant differences between the KSA compared with BAT at level of significance of 0.05, because 0.674 > 0.05. Similarly, KSA as compared with WSA-MP, ACO, PSO, all have their p-values greater than the level of significance. This indicates that there is not statistically significant differences between KSA compared with WSA-MP, ACO, PSO and BAT.

9 Conclusion

The KSA has its own advantages in feature selection in classification. Compared with meta-heuristic algorithms, classification accuracy of KSA is comparable to ACO, BAT, WSA-MP and PSO. This suggests that the initial parameters that were chosen in KSA guarantee good solutions that is comparable to other meta-heuristic search methods on feature selection. The future work for KSA is to develop new versions of KSA with modification and enhancement of code for feature selection in classification.

Key Terminology and Definitions

Parameter tuning refers to a technique that helps with efficient exploration of search space and adaptability to different problems. The advantage of parameter tuning is that it helps assign different weighting parameters to search problems in order to find the best parameter that fits a problem.

Feature selection is defined as the process of selecting a subset of relevant features (e.g., attributes, variables, predictors) that is used in model formulation.

Feature selection in classification reduces the input variables (or attributes etc.) for processing and analysis in order to find the most meaningful inputs.

Recurrent neural network (RNN) is a discriminative model that has also been used as a generative model where "output" results from a model represent the predicted input data. When an RNN is used as a discriminative model, the output result from the model is assigned a label, which is associated with an input data.

Long short-term memory (LSTM) enables networks to remember inputs for a long time using a memory cell that acts like an accumulator, which "has a connection to itself at the next time step (iteration) and has a weight, so it copies its own real-valued state and temporal weights." But this "self-connection is multiplicatively gated by another unit that learns to decide when to clear the content of the memory." "LSTM networks have subsequently proved to be more effective, especially when they have several layers for each time step."

References

Aamodt, T. (2015). Predicting stock markets with neural networks: A comparative study. Master's Thesis.

Abd-Alsabour, N., Randall, M., & Lewis, A. (2012). Investigating the effect of fixing the subset length using ant colony optimization algorithms for feature subset selection problems. In *2012 13th International Conference on Parallel and Distributed Computing, Applications and Technologies* (pp. 733–738). IEEE.

Abdel-Hamid, O., Deng, L., & Yu. D. (2013). Exploring convolutional neural network structures and optimization for speech recognition. In *Interspeech* (Vol. 11, pp. 73–5).

Abdel-Hamid, O., Mohamed, A., Jiang, H., & Penn, G. (2012). Applying convolutional neural networks concepts to hybrid NN-HMM model for speech recognition. In *2012 IEEE international Conference on Acoustics, Speech and Signal Processing (ICASSP)* (pp. 4277–4280). IEEE.

Aboudi, N. E., & Benhlima, L. (2016). Review on wrapper feature selection approaches. In *2016 International Conference on Engineering & MIS (ICEMIS)* (pp. 1–5). IEEE.

Agbehadji, I. E. (2011). Solution to the travel salesman problem, using omicron genetic algorithm. Case study: tour of national health insurance schemes in the Brong Ahafo region of Ghana. Online Master's Thesis.

Agbehadji, I. E., Millham, R., & Fong, S. (2016). Wolf search algorithm for numeric association rule mining. In *2016 IEEE International Conference on Cloud Computing and Big Data Analysis (ICCCBDA 2016)*. Chengdu, China.

Agbehadji, I. E., Millham, R., & Fong, S. (2016). Kestrel-based search algorithm for association rule mining and classification of frequently changed items. In: *IEEE International Conference on Computational Intelligence and Communication Networks*, Dehadrun, India. 10.1109/CICN.2016.76.

Al-Ani, A., & Al-Sukker, A. (2006). Effect of feature and channel selection on EEG classification. In *2006 International Conference of the IEEE Engineering in Medicine and Biology Society* (pp. 2171–2174). IEEE.

Al-Ani, A. (2007). Ant colony optimization for feature subset selection. *World Academy of Science, Engineering and Technology International Journal of Computer, Electrical, Automation, Control and Information Engineering, 1*(4).

Almuallim, H., & Dietterich, T. G. (1994). Learning boolean concepts in the presence of many irrelevant features. *Artificial Intelligence, 69*(1–2), 279–305.

Batres-Estrada, G. (2015). Deep learning for multivariate financial time series.

Ben-Bassat, M. (1982). Pattern recognition and reduction of dimensionality. In P. R. Krishnaiah & L. N. Kanal (Eds.), *Handbook of statistics-II* (pp. 773–791), North Holland.

Berka, P., & Rauch, J. (2010). Machine learning and association rules. University of Economics

Binh, T. Z. M., & Bing, X. (2014). *Overview of particle swarm optimisation for feature selection in classification* (pp. 605–617). Berlin: Springer International Publishing.

Bishop, C. M. (2006). Pattern recognition and machine learning. Available on http://users.isr.ist.utl.pt/~wurmd/Livros/school/Bishop%20-%20Pattern%20Recognition%20And%20Machine%20Learning%20-%20Springer%20%202006.pdf.

Blum, A. L., & Langley, P. (1997). Selection of relevant features and examples in machine learning. *Artificial Intelligence, 97,* 245–271.

Boser, B. E., Guyon, I. M., & Vapnik, V. N. (1992). A training algorithm for optimal margin classier. http://w.svms.org/training/BOGV92.pdf.

Dorigo M., & Cambardella, L. M. (1997). Ant colony system: A cooperative learning approach to traveling salesman problem. *IEEE Transactions on Evolutionary Computation, 1* (1), 53–66.

Collobert, R., Weston, J., Bottou, L., Karlen, M., Kavukcuoglu, K., & Kuksa, P. (2011). Natural language processing (almost) from scratch. *Journal of Machine Learning Research, 12,* 2493–2537.

Cui, X., Gao, J., & Potok, T. E. (2006). A flocking based algorithm for document clustering analysis. *Journal of Systems Architecture, 52*(8–9), 505–515.

Dash, M., & Liu, H. (1997). Feature selection for classification, intelligent data analysis. *1*, 131–156.

Deng, L. (2011). An overview of deep-structured learning for information processing. In *Proceedings of Asian-Pacific Signal & Information Processing Annual Summit and Conference (APSIPA-ASC).*

Deng, L. (2012). Three classes of deep learning architectures and their applications: A tutorial survey. *APSIPA Transactions on Signal and Information Processing*

Deng, L., & Chen, J. (2014). Sequence classification using the high-level features extracted from deep neural networks. In *Proceedings of International Conference on Acoustics Speech and Signal Processing (ICASSP).*

Deng, L., & Yu, D. (2013). Deep learning: Methods and applications. *Foundations and trends in signal processing, 7*(3–4), 197–387.

Elisseeff, A., & Guyon, I. (2003). An introduction to variable and feature selection. *Journal of Machine Learning Research, 3*(2003), 1157–1182.

Englert, P., Paraschos, A., Peters, J., & Deisenroth, M. P. (2013). Probabilistic model-based imitation learning. http://www.ias.tu-darmstadt.de/uploads/Publications/Englert_ABJ_2013.pdf.

Ferchichi, S. E., Laabidi, K., Zidi, S., & Maouche, S. (2009). Feature Selection using an SVM learning machine. In *2009 3rd International Conference on Signals, Circuits and Systems (SCS)* (pp. 1–6). IEEE.

Fong, S., Yang, X.-S., & Deb, S. (2013). Swarm search for feature selection in classification. In *2013 IEEE 16th International Conference on Computational Science and Engineering.*

García, S., Fernández, A., Benítez, A. D., & Herrera, F. (2007). Statistical comparisons by means of non-parametric tests: A case study on genetic based machine learning. http://www.lsi.us.es/redmidas/CEDI07/%5B9%5D.pdf.

Graves, A., & Jaitly, N. (2014). Towards end-to-end speech recognition with recurrent neural networks. In *International Conference on Machine Learning* (pp. 1764–1772).

Hall, M. A. (2000). Correlation-based feature selection for discrete and numeric class machine learning. In *Proceedings of 17th International Conference on Machine Learning* (pp. 359–366).

Holland, J. (1975). *Adaptation in natural and artificial systems.* Ann Arbor, MI: University of Michigan Press.

Honkavaara, J., Koivula, M., Korpimäki, E., Siitari, H., & Viitala, J. (2002). Ultraviolet vision and foraging in terrestrial vertebrates. https://projects.ncsu.edu/cals/course/zo501/Readings/UV%20Vision%20in%20Birds.pdf.

Kennedy, J., & Eberhart, R. C. (1995). Particle swarm optimization. In *Proceedimgs of IEEE International Conference on Neural Networks* (pp. 1942–1948), Piscataway, NJ.
Kim, J. W. (2013). Classification with deep belief networks. Available on https://www.ki.tu-berlin.de/fileadmin/fg135/publikationen/Hebbo_2013_CDB.pdf.
Kohavi, R., & John, G. H. (1996). Wrappers for feature subset selection. *Artificial Intelligence, 97*(1–2), 273–324.
Krause, J., Cordeiro, J., Parpinelli, R. S., & Lopes, H. S. (2013).A survey of swarm algorithms applied to discrete optimization problems.
Krizhevsky, A., Sutskever, I., & Hinton, G.E. (2012). Imagenet classification with deep convolutional neural networks. In *Proceedings of the Twenty-Sixth Annual Conference on Neural Information Processing Systems* (pp. 1097–1105). Lake Tahoe, NY, USA, 3–8 December 2012.
Kumar, R. (2015). Grey wolf optimizer (GWO).
Kumar, V., & Minz, S. (2014). Feature selection: A literature review. *Smart Computing Review, 4*(3).
Le, Q. V. (2015). A tutorial on deep learning part 1: Nonlinear classifiers and the backpropagation algorithm.
LeCun, Y., Bengio, Y., & Hinton, G. (2015). Review: Deep learning. *Nature, 521*(7553), 436–444.
LeCun, Y., Bottou, L., Bengio, Y., & Haffner, P. (1998). Gradient-based learning applied to document recognition. *Proceedings of the IEEE, 86,* 2278–2324.
Lee, H., Grosse, R., Ranganath, R. & Ng, A. Y. (2009). Convolutional deep belief networks for scalable unsupervised learning of hierarchical representations. In *ICML.*
Li. D. (2013). Three classes of deep learning architectures and their applications: A tutorial survey. research.microsoft.com.
Li, J., Fong, S., Wong, R. K., Millham, R., & Wong, K. K. L. (2017). Elitist binary wolf search algorithm for heuristic feature selection in high-dimensional bioinformatics datasets. *Scientific Reports, 7*(1), 1–14.
Liang, J., Wang, F., Dang, C., & Qian, Y. (2012). An efficient rough feature selection algorithm with a multi-granulation view. *International Journal of Approximate Reasoning, 53*(6), 912–926.
Lin, C.-J. (2006). Support vector machines: status and challenges. Available on https://www.csie.ntu.edu.tw/~cjlin/talks/caltech.pdf.
Liu, H., & Yu, L. (2005). Towards integrating feature selection algorithms for classification and clustering. *IEEE Transactions on Knowledge and Data Engineering, 17*(4).
Longbottom, C, & Bamforth, R. (2013). Optimising the data warehouse. *Dealing with large volumes of mixed data to give better business insights.* Quocirca.
Mafarja, M., & Mirjalili, S. (2018). Whale optimization approaches for wrapper feature selection. *Applied Soft Computing, 62,* 441–453.
Marcus, G. (2018). Deep learning: A critical appraisal. https://arxiv.org/abs/1801.00631.
Marill, D. G. T. (1963). On the effectiveness of receptors in recognition systems. *IEEE Transactions on Information Theory, 9*(1), 11–17.
Patel, A. B., Nguyen, T., & Baraniuk, R. G. (2015). A probabilistic theory of deep learning. arXiv preprint arXiv:1504.00641.
Qui, C. (2017). Bare bones particle swarm optimization with adaptive chaotic jump for feature selection in classification. *International Journal of Computational Intelligence Systems, 11*(2018), 1–14.
Sainath, T., Mohamed, A., Kingsbury, B., & Ramabhadran, B. (2013). Deep convolutional neural networks for LVCSR. In *2013 IEEE International Conference on Acoustics, Speech and Signal Processing* (pp. 8614–8618). IEEE.
Shrubb, M. (1982). The hunting behaviour of some farmland Kestrels. *Bird Study, 29,* 121–128.
Siripurapu, A. (2015). *Convolutional networks for stock trading.* Stanford University Department of Computer Science, Course Project Reports
Sohangir, S., Wang, D., Pomeranets, A., & Khoshgoftaar, T. M. (2018). Big data: Deep learning for financial sentiment analysis. *Journal of Big Data, 5*(1), 3.
Spencer, R. L. (2002). Introduction to Matlab.

Stützle, T., & Dorigo, M. (2002). The ant colony optimization metaheuristic: algorithms, applications, and advances. In F. Glover & G. Kochenberger (Eds.), *Handbook of metaheuristics*. Norwell, MA: Kluwer Academic Publishers.

Tang, R., Fong, S., Yang, X.-S., & Deb, S. (2012). Wolf search algorithm with ephemeral memory.

Tian, Z., & Fong, S. (2016). Survey of meta-heuristic algorithms for deep learning training. *Optimization algorithms—methods and applications.*

Uncu, O., & Turksen, I. B. (2007). A novel feature selection approach: Combining feature wrappers and filters. *Information Sciences, 177*(2007), 449–466.

Unler, A., & Murat, A. (2010). A discrete particle swarm optimization method for feature selection in binary classification problems. *European Journal of Operational Research, 206*(3), 528–539.

Varland, D. E. (1991). Behavior and ecology of post-fledging American Kestrels. Retrospective Theses and Dissertations Paper 9784.

Vlachos, C, Bakaloudis, D., Chatzinikos, E., Papadopoulos, T., & Tsalagas, D. (2003). Aerial hunting behaviour of the lesser Kestrel falco naumanni during the breeding season in thessaly (Greece). *Acta Ornithologica, 38*(2), 129–134.

Waad, B., Ghazi, B. M., & Mohamed, L. (2013). On the effect of search strategies on wrapper feature selection in credit scoring. In *2013 International Conference on Control, Decision and Information Technologies (CoDIT)* (pp. 218–223). IEEE.

Weston, J., Chopra, S., & Adams, K. (2014). # tagspace: semantic embeddings from Hashtags. In Proceedings of the 2014 Conference on Empirical Methods in Natural Language Processing (EMNLP) (pp. 1822–1827).

Whitney, A. W. (1971). A direct method of nonparametric measurement selection. *IEEE Transactions on Computers, C-20*(9), 1100–1103.

Xue, B., Bing, W. N., & Zhang, M. (2014). Particle swarm optimisation for feature selection in classification: Novel initialisation and updating mechanisms. *Applied Soft Computing, 18,* 261–276.

Zar, J. H. (1999). *Biostatistical analysis*. Prentice Hall.

Richard Millham is currently an Associate Professor at Durban University of Technology in Durban, South Africa. After thirteen years of industrial experience, he switched to academe and has worked at universities in Ghana, South Sudan, Scotland, and Bahamas. His research interests include software evolution, aspects of cloud computing with m-interaction, big data, data streaming, fog and edge analytics, and aspects of the Internet of Things. He is a Chartered Engineer (UK), a Chartered Engineer Assessor and Senior Member of IEEE.

Israel Edem Agbehadji graduated from the Catholic University college of Ghana with B.Sc. Computer Science in 2007, M.Sc. Industrial Mathematics from the Kwame Nkrumah University of Science and Technology in 2011 and Ph.D. Information Technology from Durban University of Technology (DUT), South Africa, in 2019. He is a member of ICT Society of DUT Research group in the Faculty of Accounting and Informatics; and *IEEE* member. He lectured undergraduate courses in both DUT, South Africa, and a private university, Ghana. Also, he supervised several undergraduate research projects. Prior to his academic career, he took up various managerial positions as the management information systems manager for National Health Insurance Scheme; the postgraduate degree program manager in a private university in Ghana. Currently, he works as a Postdoctoral Research Fellow, DUT-South Africa, on joint collaboration research project between South Africa and South Korea. His research interests include big data analytics, Internet of Things (IoT), fog computing and optimization algorithms.

Hongji Yang graduated with a Ph.D. in Software Engineering from Durham University, England with his M.Sc. and B.Sc. in Computer Science completed at Jilin University in China. With over 400 publications, he is full professor at the University of Leicester in England. Prof Yang has been an IEEE Computer Society Golden Core Member since 2010, an EPSRC peer review college member since 2003, and Editor in Chief of the International Journal of Creative Computing.

Chapter 3
Data Stream Mining in Fog Computing Environment with Feature Selection Using Ensemble of Swarm Search Algorithms

Simon Fong, Tengyue Li, and Sabah Mohammed

1 Introduction

Generally, fog computing is also referred to as fog networking or fogging. In principle, fog computing is an extension of the cloud computing framework. The difference is that fog computing is at the edge of a network (that is the primary location of devices) to allow timely interaction with sensor networks which was earlier handled by the cloud computing framework. Subsequently, the data analytics workload at cloud computing platform could be delegated to nodes at the edge of a network instead of the Central cloud server. Hence, fog computing is a layer found between a sensor-enabled device and the cloud computing framework. The basis is to provide timely and accurate processing of data stream from sensors enabled devices.

The framework of fog computing consists of four basic components namely terminal, platform, storage and security. The basic components enable data preprocessing and analyzing patterns from the incoming data streams before being transferred to the cloud computing framework for historical analysis and storage. Thus, fog computing is better suited for edge network, where data can diminish quickly, as it reduces data transfer to the cloud computing framework and minimize data

S. Fong (✉)
Department of Computer Science, University of Macau, Taipa, Macau SAR
e-mail: ccfong@umac.mo

T. Li
Center of Big Data and Cloud Computing, Zhuhai Institute of Advanced Technology, Chinese Academy of Science, Zhuhai, China
e-mail: litengyue2018@gmail.com

S. Mohammed
Department of Computer Science, Lakehead University, Thunder Bay, Canada
e-mail: sabah.mohammed@lakeheadu.ca

S. J. Fong and R. C. Millham (eds.), *Bio-inspired Algorithms for Data Streaming and Visualization, Big Data Management, and Fog Computing*, Springer Tracts in Nature-Inspired Computing,
https://doi.org/10.1007/978-981-15-6695-0_3

analytics latency. Consequently, this increases the efficiency of the IoT operation in this current dispensation of unprecedented amount of data stream every second from several sensor-enabled devices. In view of this, it is significant to consider the speed, efficiency and accuracy of stream data mining algorithms that support edge intelligence.

Practically, fog computing architecture has different kinds of components and functions. Fog computing gateways accept the data from the end devices, such as routers and different switching equipment graphically distributed eventually global public or private cloud services and servers. Security network plays a vital role in fog computing, the virtual firewalls are necessary to design. In conclusion, fog computing provides logical structure and a model, that is used to solve exabytes of data generated by IoT end devices. It will help process the data closure to the point of origin and duly solving the challenges of exploding data volume, variety and velocity. It is a benefit for lowering the response time through saving bandwidth to eliminate sending data to the cloud. Ultimately 'time-sensitive data' transfers and analyzes close to where it is generated instead of sending gigantic data to the cloud. Finally, the fog computing has expanded the network computing model of fog computing, extending network computing from the centre of the network to the edge of the network, and has been more widely used in various services, which contributed to the users easily achieving efficiently insights, leading to benefit for business agility, effectively services and improved data.

The increasingly creative and multifunctional smart devices such as sensors and smartphones have been promoting the fast development of data streaming applications, such as event monitoring, interactive device. The numerous data streams produced by these data streaming applications, with the rapid development of IoT, promote applications for high-level analysis over the substantial sensor data streams, IoT devices generate data continuously, sending and analysis should be fast enough to manage this data. For instance, when the gas value in a building is rapidly approaching to the acceptable limit, thoughtful action must be taken almost immediately. A new computing model is necessary for solving the volume, variety and velocity of IoT data. Minimizing latency is necessary because the end to end delay may not match the requirement of numerous data streaming applications. For example, the augmented reality applications typically need approximately 10 ms for a response, but it is too difficult to achieve hundreds of millisecond latency. The millisecond latency is important when the manufacturing industry close their lines suddenly or in the case of electronic devices restoration. Because the fog calculation is closer to the ground than cloud computing, this latency time is reduced. Specifically, their position in the network topology is different. Analyzing data close to the device is a benefit for averting disaster. Fundamentally providing low latency network connections between devices is crucial for quick response times. Focusing on converse network bandwidth is a critical action that should be used; it is not practical to transport a huge amount of data from thousands or hundreds of thousands of edge devices to the cloud. IoT data is rapidly used for decisions in citizen secure and essential infrastructure. IoT data pay attention to both in transit and at rest to deal with security issues. Collecting and securing data across a wide range of geographic areas in a variety of circumstances. IoT devices can be distributed over hundreds or more square miles. Devices deployed in bad conditions such as roadways, railways, a set of public equipment. Extremely time-sensitive decisions should be made closer to the things producing the data. Traditional cloud computing is not a match for these requirements.

In this chapter, we consider a fog computing scenario where air/gas samples are collected from sensors and the model that is built by the algorithm(s) in the form of a decision tree classifies what type of gas it is. Decision tree is a machine learning model where tree branches are extracted into useful predicate-type of decision rules which are coded as a set of logics into embedded devices. In addition, correlation-based feature selection algorithm, and traditional search methods with ensemble of swarm search methods are integrated into the data stream mining algorithm as preprocessing mechanism. In this experiment, we take into account the real-time constraint and capability requirements of processing devices to select the best algorithm suitable for data stream. When the accuracy is similar, there are three data stream mining performance criteria that are put forward. Recovery ability is used to judge stable data. Accumulate degree is used to calculate how many times the curve is over the quartile line. Successive time is to calculate the curve over some threshold from the beginning until it drops below the threshold.

2 Proposed Methodology

In this part, this case emulates an emergency rescue service system based on Internet of Things, which focuses on verifying the feasibility and effectiveness of applying fog computing in emergency service. The IoT is used in the field of urban emergency rescue service, which effectively solves the existing weaknesses including nontimely warnings and dispatching of response services. It is based on requirements of the emergency rescue service system and technical peculiarity of IoT, there is an innovative concept called Internet of Breath (IoB) that makes a contribution to fire-and-rescue (FRS) operations. It provided real-time information transmission about air quality data and the situation information to the firefighters in the field.

2.1 IoT Emergency Service

The Internet of Breath program suggests installing a network of gas sensors at a wireless sensor network to complement an existing urban automatic fire detection system (SADI). The main function of IoB is to recognize abnormal gas by collecting and analyzing different types of gas data and CO_2. Real-time monitoring of the variation of gas concentrations triggers off some emergency alarm, as well as providing comprehensive air quality information at the proximity as an early warning. IoB provides data analytic results such as fire severity, estimated degree of damage, duration and fire direction through collected continuous air data that have been collected and analyzed using machine learning; meanwhile it will transmit promptly the real-time information of the fire field. Therefore, combining traditional IoT architecture with an emergency rescue system, four layers of IoT emergency rescue service system architecture is designed as shown in Fig. 1. They are the perception layer, network layer, support layer and application layer, respectively. The first layer is composed of different kinds of sensors used for collecting data and distinguishing air samples.

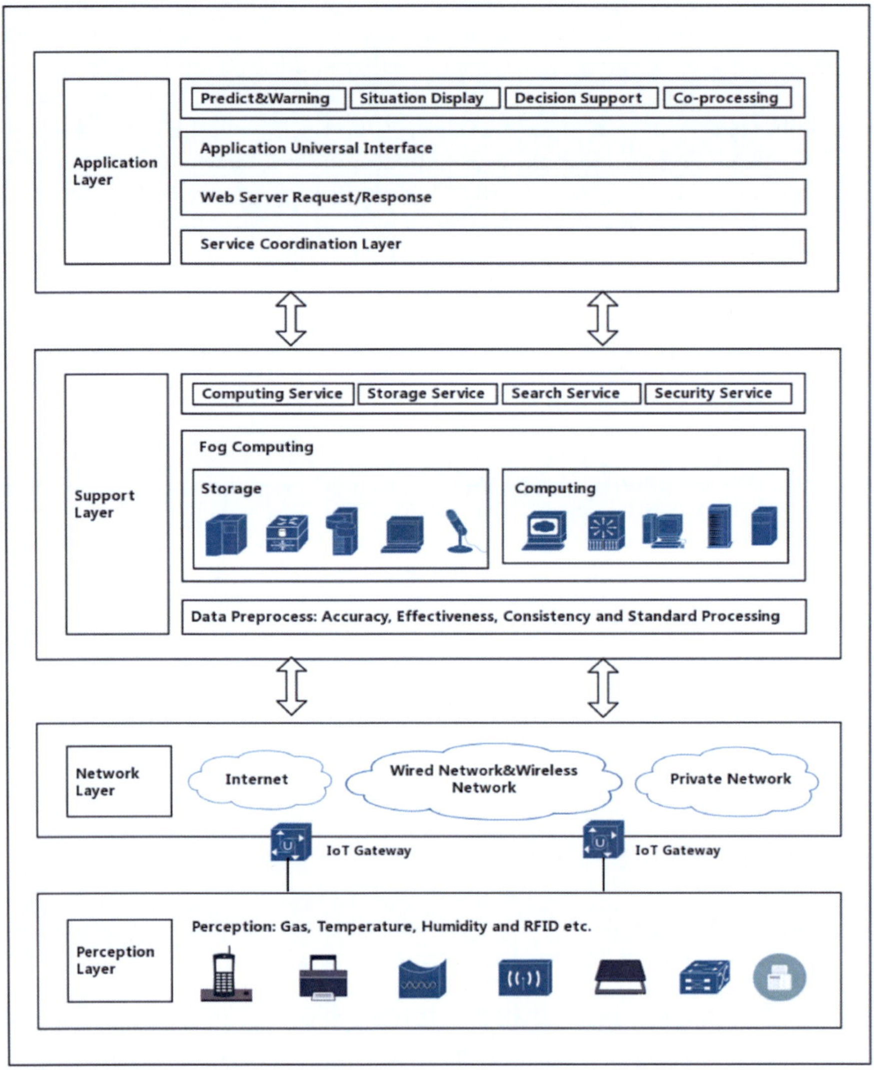

Fig. 1 IoT emergency rescue service system architecture

The function of the network layer offers network connectivity support. The third layer is based on fog computing and divided into data preprocessing, fog computing and basic service, respectively, which provide basic computing, storage and service. The top layer is the application layer that is mainly used to sharing and reusing the services and messages supporting the emergency rescue system. Scheduling strategy is to achieve various functional department cooperation by intellectual services.

IoB can approximate detect and count the occupancy of humans in each room with a CO_2 sensor along with an existing fire detection system. Whether a room was occupied or empty would be known just prior to a break out of fire. Of course, the people might have left already in case of fire; but this information would still be useful in planning for fire rescue. Traditionally a fire detector, which connects to water sprinkler water when the thermal fuse is disconnected, only signals on when a fire was already developed to a certain intensity. Early information about how the fire developed through the analysis of the increase of CO_2 concentration helps in estimating the growth, the intensity and spread of fire. Valuable measurements supported by IoB include ethanol, ethylene, ammonia, acetaldehyde, acetone and toluene. Especially, the new sensor consists of 16 chemical sensors utilized in simulations, which measures the ambient background and wave of the six gas concentration. Through the analysis, it estimates how many people are present at the fire field instead of fire marshal relying on post-incident statistics. To collect, analyze and store data in real-time offers a dynamic view on how, where and when the fire broke up; monitoring its development emergency rescue work could be improved from knowing such real-time information. With IoB, it is possible to transform from passive emergency rescue to active early warning and disaster prevention if the fire could be pinpointed and monitored early (Fig. 2).

In the data streaming, the mean accuracy plays a vital role in measuring the performance of the algorithms, however, sometimes the mean accuracy cannot tell you all about the whole data stream mining.

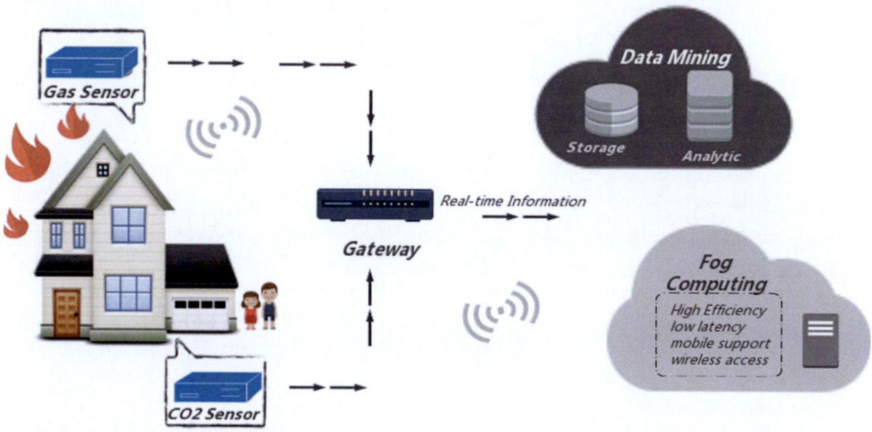

Fig. 2 The architecture of IoB real-time information transmission

For example, two algorithms have similar mean accuracy, the curve of the first algorithm is very smooth, and another algorithm's curve is going up and down dramatically. Hence, we propose some criteria that can measure the performance of data stream mining when the compared algorithms have a similar mean accuracy.

2.2 *Data Stream Mining Performance Criteria*

2.2.1 Recovery Ability

The first criteria that we would like to propose is recovery ability. Data stream mining algorithms sometimes may not be stable enough to maintain good performance, so there may occur some big drops during the whole data stream mining. We want to measure the ability to recover from the drop that we have mentioned. Before we define the recovery ability, we need to figure out how to define a drop by a specific formula. We would like to use quantile to help us. At first, we calculate the halves, thirds and quarters quantile for the whole accuracy during the process. When the curve of the accuracy decreases to the value of Q1, then we record the point as the start drop when the curve successively drops to a threshold (in this thesis, we define the Q3-30% as the threshold, this can be adjusted by the users based on different cases), we think it is a selected drop that will be used to calculate the recovery time. When the curve back to the Q1, we record the stop drop point of the curve. The time of start drop point to the stop drop point is called the Q1 recovery time. If the curve has more than one drop, we choose the curve with the longest recovery time. The Q2 and Q3 recovery time are calculated in the same way. For each recovery time for the same data mining algorithm, we allocate different weights for different quantiles. The weight for halves, thirds and quarters quantile is 0.2, 0.3 and 0.5, respectively. The reason why we define them in this way is that we think the drop for Q3 is the most important and Q1 is the least important, the algorithm can recover to high accuracy in quite a short time means better performance.

Quantile: Q1 (25%) Q2 (50%) Q3 (75%)

Q n (n = 1, 2, 3) recovery time: a drop crush from Qn to a threshold (Q3-30%)

Recovery Ability = 1/(Q1 recovery time × w1 + Q2 recovery time × w2 + Q3 recovery time × w3)

$w1 + w2 + w3 = 1$

Figure 3 shows an example, the curve drops from the 5.7564 and recover at 8.7516, and it reached the threshold of Q3-30%, so the recovery time for Q1 is 2.9952.

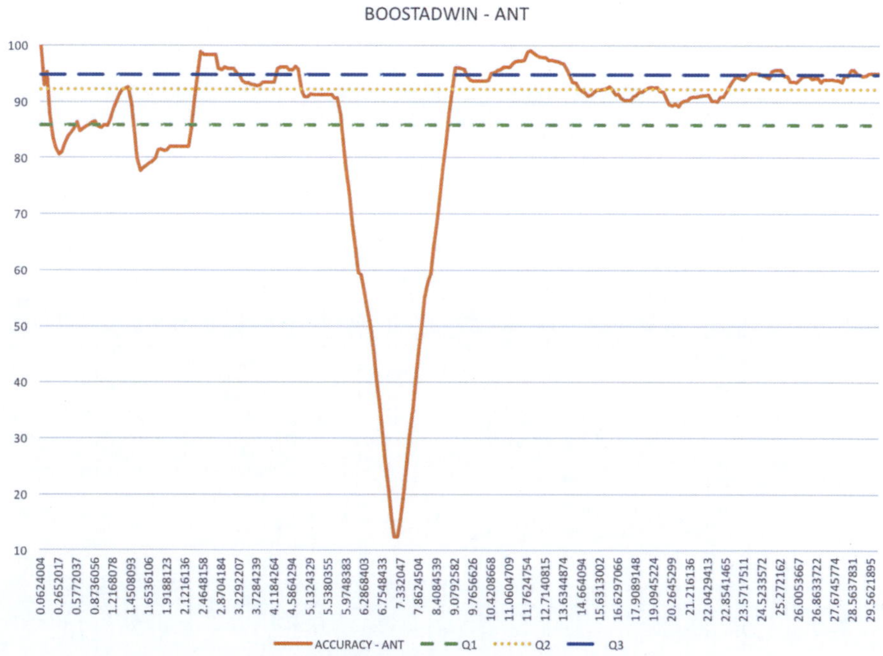

Fig. 3 The recovery ability of BOOSTADWIN—ANT

2.2.2 Fluctuating Degree

The second criterion that we would like to propose is the Fluctuating Degree. It is defined for measuring the stability of the data stream mining algorithms. For data stream mining algorithms, the accuracy may always go up and down, some algorithms may increase to very high accuracy and drop to an extremely low accuracy several times during the whole data mining process. Even the algorithm has such unstable performance; the average accuracy for this algorithm may be close to some other stable ones. In order to solve this problem, we propose this measurement. This criterion also uses quantile as the threshold and we use the Q1 as an example. We calculate the accumulated times for each quantile, and the accumulated times are how many times the curve is over the quantile lines. When the curve increases to the value of Q1, it is considered as the start of the candidate curve, and when the curve decreases to Q1, we think it is the end of the candidate curve. From the start to the end of the candidate curve, the whole curve should be over the Q1 line. The Q1 accumulate times is how many candidate curves during the whole data mining process. The weight is similar to the recovery ability (Fig. 4).

Quantile: Q1 (25%) Q2 (50%) Q3 (75%)

Qn accumulate times: how many times accuracy over the Qn (n = 1, 2, 3)

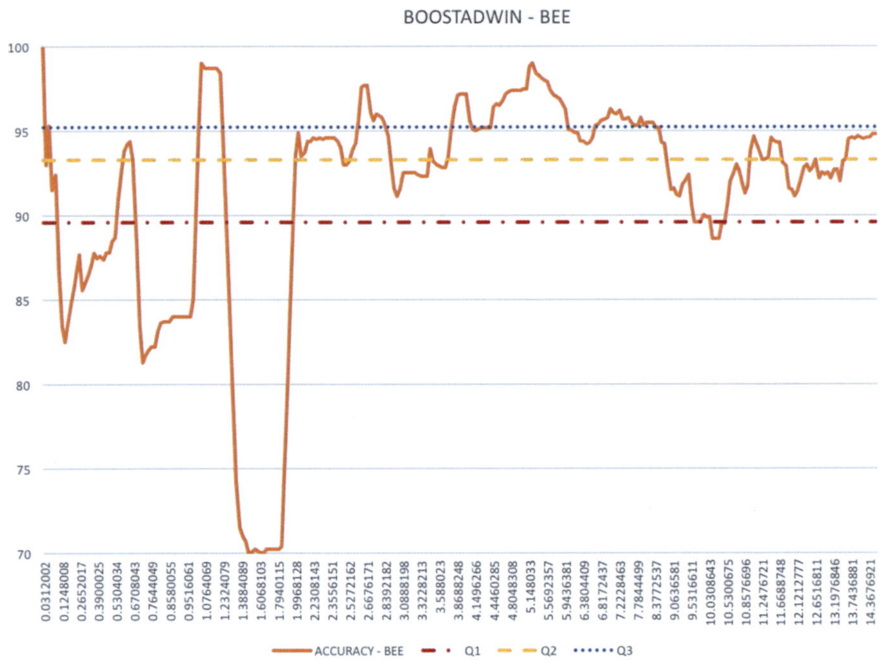

Fig. 4 The recovery ability of BOOSTADWIN—BEE

Fluctuating Degree = accumulate times Q1 × W1 + accumulate times Q2 × W2 + accumulate times Q3 × W3

w1 + w2 + w3 = 1

2.2.3 Successive Time

The third criterion is the successive time. Sometimes, the users may care about how long the good service that the algorithms can provide from the beginning maybe, so we propose a new criterion to measure it. This criterion is also based on the compared data stream mining algorithms that have close average accuracy. Like the other two criteria, we use the quartile as threshold and use Q1 as an example. The successive time for Q1 is the curve that has the accuracy over Q1 from the beginning until it decreased to the value of Q1 line. Then we use the weight times them to form the successive service quality. The sum of each weight is 1, the Q1 is also the least important, and the Q3 is the most important. The reason is the same as the other two algorithms. For example, the successive time for Q3 is about 0.0312002. For Q2 is 0.2184014. For Q1 is 0.2496016. And we set the weight W1 = 0.2, W2 = 0.3 and W3 = 0.5 (Fig. 5).

Quantile: Q1 (25%) Q2 (50%) Q3 (75%)

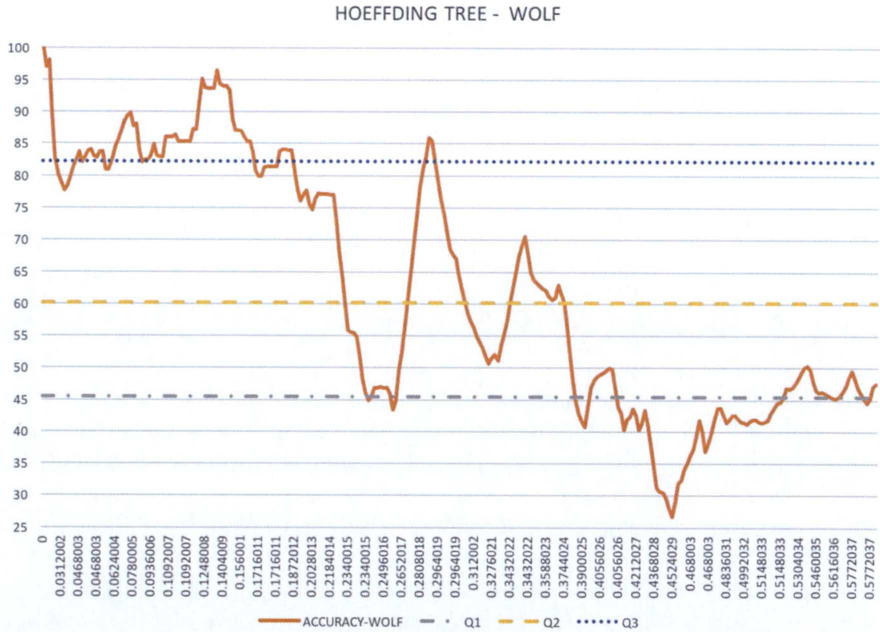

Fig. 5 The recovery ability of HOEFFDING TREE—WOLF

Threshold: Q1, Q2, Q3

Successive time = time for curve over Qn (n = 1, 2, 3)

Successive service quality = successive time Q1 × W1 + successive time Q2 × W2 + successive time Q3 × W3

w1 + w2 + w3 = 1

In outlier detection, we can use many detection indices to find out those abnormal values, like LOF value or Mahalanobis. What we should not ignore is that these indices are all calculated after a mathematical operation. That means these values are generated from some complicated formula. Sometimes we should aware that another outlier detection direction, which called a statistical operation. These two operations do not have conflicts with each other.

3 Data Stream Mining Performance Criteria

3.1 Experiment Setup

The example setup of how the gas sensor data are collected, which is shown in computer-supervised continuous flow system. In the beginning, there are three pressurized gas cylinders, zero grade dry air, Odorant 1 and Odorant 2, respectively, which go through the mass flow controller, the data are collected by the sensor that is 60-ml volume test chamber on the electronic board. Total vapour flow through the test chamber: 200 ml/min. The temperature control via the heater voltage. The data acquisition via DACC board.

In the room condition include wind direction, wind speed and room temperature, all mentioned during the entire measurement process, air inlet at room conditions, the red circle is the chemical source, the whole length is 2.5 m, the position label is P1 = 0.25, P2 = 0.5, P3 = 0.98, P4 = 1.18, P5 = 1.40 and P6 = 1.45, respectively. The data is increasing tremendously and continuously. In gas monitoring, data generated from chemical sensors would need to be collected frequently. It is needed to know constantly whether room condition is safe by recognizing any drift compensation in a discrimination task at different levels of concentrations. The outlet is 12 V, 4.8 A, 1500–4400 rpm.

A simulation experiment is designed to test the possibilities of using decision tree models by two types of algorithms for supporting the fog data analytics on IoB. The fog analytics is supposed to have integrated preprocessing with decision tree model. Searching for a suitable algorithm applied to fog computing environment is the experimentation objective. The experiment setting is related to the fog analytics which is to be installed at the gas sensor gateway where the hardware device collects gas quality data continuously. The edge analysis is powered by a decision tree model that is built from crunching over the continuous data stream. Located at the edge node, the decision tree could be built by any chosen algorithm that is found to be most suitable for the gas data streaming patterns. Owing to different types of decision tree algorithm model (traditional vs data stream mining), and different search algorithms are available in the feature selection, the algorithms are put under test for evaluation in fog computing and IoT emergency rescue service environment. Performance indicators like accuracy, Kappa and time cost are measured.

Fog computing is different from cloud computing in data analytics. However, because fog computing is relatively a new area, little is known on how best to use different data mining methods for different fog scenarios. Fog computing usually does not need to load the whole dataset into the data mining model, unlike big intelligence where it is meant to be obtained from big historical data, each time, monthly intelligence reports are generated, time cost is high in rebuilding the model. However, the data stream mining for fog computing only reads the data once and incrementally it updates/refreshes the model at the network edge.

For strengthening the data mining model, data preprocessing is significant at the beginning of model induction. The usual process begins from data cleaning, data integration, data normalization and data replacement and feature selection. In this simulation experiment, focusing on feature selection is tested in combining the conventional decision tree algorithm (C4.5) with data mining decision tree algorithm (HT). There are several types of feature selection search methods being tested.

In the experiment, the dataset is named Gas Sensor Array Drift, which includes 13,910 data instances of measurements are collected from 16 chemical sensors. The data were observed to have drift compensation in a classification task that groups the data into one of the six gases at certain concentrations levels. This dataset used for the experiment benefit to simulate IoB environment, where it might be possible to deal with data that are infested with concept drift. The six classes belong to the following six gas types, such as Ammonia, Acetaldehyde, Acetone, Ethylene, Ethanol and Toluene. IoB target is to detect and recognize harmful gas in a fire scene. This dataset is closer to the dataset of a fire field where different types of chemical gases may be shown.

The simulation platform is WEKA and MOA—Massive Online Analysis which are machine learning (as well as data mining) and data stream mining benchmarking software programs by the University of Waikato, New Zealand. The hardware platform is MacBook Pro, i7-CPU and 16 MB RAM.

3.2 Gas Sensor Classification Results

There are two steps in the experiment. Firstly, comparing the traditional decision tree algorithm C4.5 with the data mining decision tree algorithm called HOEFFDING tree following with the original dataset and with feature selection algorithm on the two classifiers. Table 1 shows the detail of comparing C4.5 with HT in the different feature selection algorithm.

It is obviously showing that the accuracy of C4.5 is higher than HT in general in Table 1: C4.5 is built with the whole dataset and training is sufficient, therefore the result is more precise. Best first search method shows the highest accuracy in C4.5, Flower and Elephant search methods are higher than the others. As a huge size of gas sensor dataset, the accuracy decreases when it is applied to C4.5. The accuracy is improving in data stream mining using HT, meanwhile, the accuracy increases with all FS search methods than original applied to HT. FS enables HT to learn and predict target, concerning on related data benefit to learning. With regards to FS in data stream mining, the Harmony algorithm is more effective than the other search method in HT of the accuracy close to 97.4397%. Harmony is better than the other search method in accuracy, Kappa and TP rate in HT. The following chart illustrates the performance comparison of HT algorithms in MOA platform combined with some feature selection search algorithms. In this experiment, HT data stream mining algorithm performance is shown here. It can evaluate different feature selection algorithms.

Figure 6 demonstrates the accuracy performance in HT, which focuses on fluctuating at different points of time. The accuracy maintains high in a range between

Table 1 Comparison in C4.5 and HOEFFDING TREE

C4.5	Accuracy	Kappa	TP Rate	FP Rate	Precision	Recall	*F*-Measure
Original	99.66	0.9771	0.997	0.027	0.997	0.997	0.997
Best first	98.9699	0.9284	0.99	0.095	0.99	0.99	0.99
PSO	99.0599	0.9365	0.991	0.062	0.991	0.991	0.991
Ant	99.0699	0.9366	0.991	0.07	0.991	0.991	0.991
Bat	99.3899	0.9587	0.994	0.047	0.994	0.994	0.994
Bee	98.6499	0.9078	0.986	0.096	0.986	0.986	0.986
Cuckoo	98.4799	0.9649	0.995	0.037	0.995	0.0995	0.9995
Elephants	99.59	0.9723	0.996	0.034	0.996	0.996	0.996
Firefly	99.6	0.973	0.996	0.031	0.996	0.996	0.996
Flower	99.4699	0.9642	0.995	0.039	0.995	0.995	0.995
GA	98.9299	0.9275	0.989	0.073	0.989	0.989	0.989
Harmony	98.4998	0.8982	0.985	0.099	0.0985	0.985	0.985
Wolf	99.2699	0.9505	0.993	0.054	0.993	0.993	0.993
Evolutionary	99.56	0.9701	0.996	0.041	0.996	0.996	0.996
HT	Accuracy	Kappa	TP Rate	FP Rate	Precision	Recall	F-Measure
Original	96.9597	0.7658	0.97	0.304	0.969	0.97	0.967
Best first	97.3997	0.7989	0.974	0.283	0.974	0.974	0.972
PSO	97.2197	0.785	0.972	0.293	0.972	0.972	0.97
Ant	97.0597	0.7714	0.971	0.306	0.97	0.971	0.968
Bat	97.2597	0.7897	0.973	0.284	0.972	0.973	0.971
Bee	97.0297	0.771	0.97	0.301	0.97	0.97	0.968
Cuckoo	97.0797	0.7782	0.971	0.285	0.97	0.971	0.969
Elephants	97.2797	0.7952	0.973	0.266	0.972	0.973	0.971
Firefly	97.0997	0.7802	0.971	0.282	0.97	0.971	0.969
Flower	97.3797	0.8052	0.974	0.249	0.973	0.974	0.972
GA	96.9197	0.7615	0.969	0.311	0.969	0.969	0.967
Harmony	97.4397	0.8087	0.974	0.25	0.974	0.974	0.973
Wolf	97.2697	0.7976	0.973	0.252	0.972	0.973	0.971
Evolutionary	97.0997	0.7791	0.971	0.286	0.97	0.971	0.969

80 and 100% and the maximum gets close to 99%. However, the average accuracy keeps at roughly 93%. It is remarkable that the accuracy has moderate growth that is eventually approaching 100% after a sharp descent at the beginning, which might be a common situation in data stream mining because of training data segments instead of the whole dataset are used. FS has a good influence on both accuracy and Kappa in HT.

Figure 7 illustrates the Kappa of HT with several FS in MOA. According to the results in the chart, we can see a clear fluctuation at the beginning and then

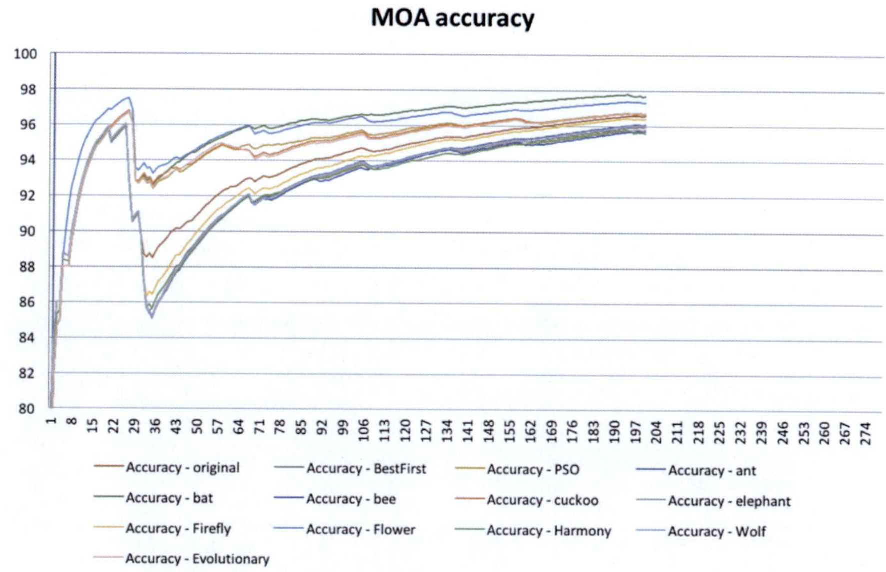

Fig. 6 Accuracy of HT with several FS in MOA

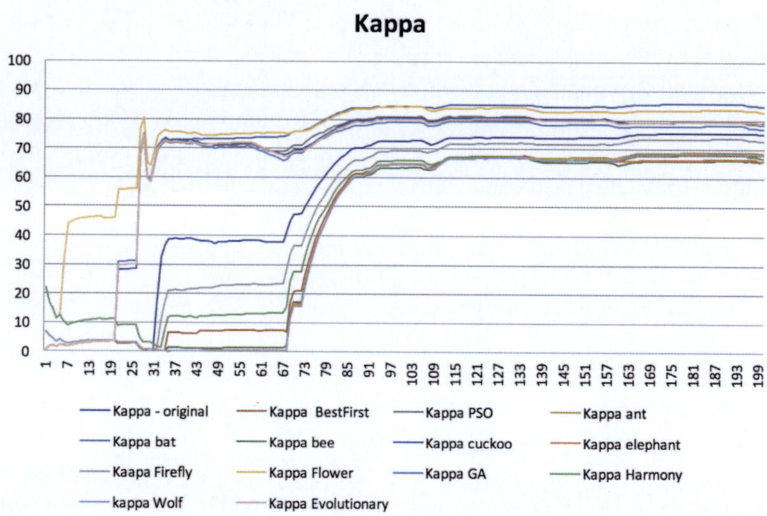

Fig. 7 Kappa of HT with several FS in MOA

eventually it becomes stable. From then on, it generally maintains an upward trend until stabilized despite some slight fluctuations; finally, the value is up to 85%. The graph indicates the Flower and the Evolutionary search methods have similar trends especially have a dramatic increase at the beginning.

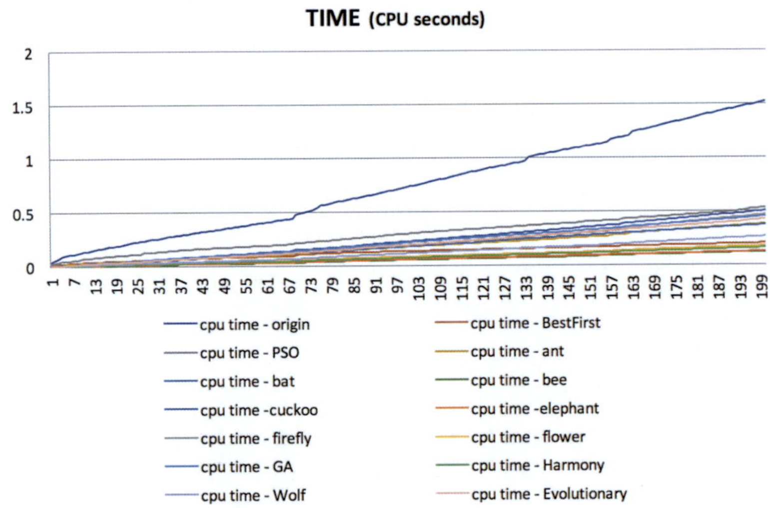

Fig. 8 Refresh time of HT with several FS in MOA

Figure 8 depicts the time performance curves; the time scales up linearly with the amount of data increases. Good news is they all scale up linearly, which is good for scalability. To utilize FS makes the time cost lower sharply, it can be seen that Harmony, Flower and Elephant are capable of decreasing the time requirement. As a result, Harmony is a good method among FS search methods in data stream mining; Harmony has better scalability following the huge amount of data arrival. In fog computing, Harmony coupled with HT is a good solution to analyze amount of data transmitted from data sensor.

3.3 Mean Accuracy Results

As we see in Fig. 9, the swarm algorithms for feature selection have very close mean accuracy when classified by the same data stream mining algorithm. This is the basic precondition for our three criteria. Based on this mean accuracy, we proposed three criteria; they are recovery ability, fluctuation degree and successive time, respectively. In the recovery ability, we calculate the reciprocal the higher is better. In the fluctuation degree, we calculate accumulation accuracy the lower is better. In the successive time, the higher is better.

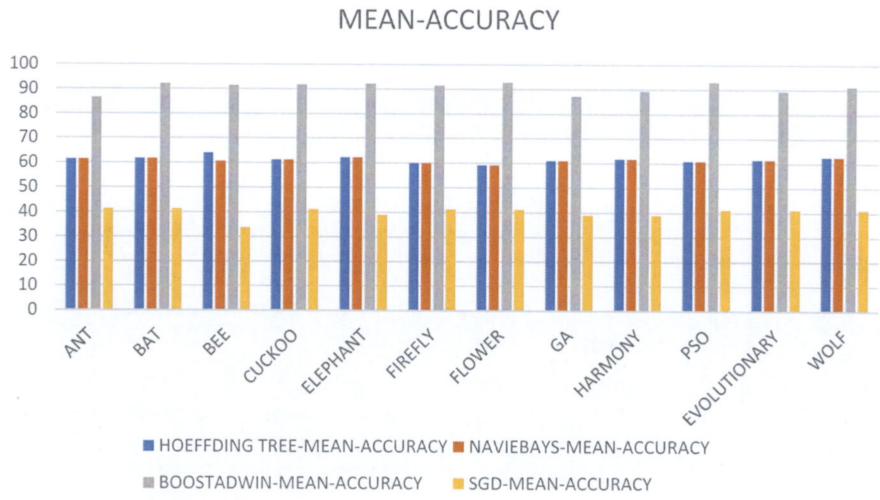

Fig. 9 The mean accuracy

3.4 Data Stream Mining Performance Criteria

3.4.1 Recovery Ability (Higher Is Better)

See Tables 2, 3, 4, 5, 6, 7, 8 and 9.

3.4.2 Accumulate Accuracy (Lower Is Better)

See Tables 10, 11, 12 and 13.

3.4.3 Successive Time (Lower Is Better)

In this experiment of the recovery time, the evolutionary algorithm is better than the others are. In the fluctuation degree, flower algorithm is better than the others. In successive time, genetic algorithm is better than the others.

4 Summary and Future Direction

Fog computing provides advantages of bringing analytics to edge intelligence. The paper shows a simulation experiment comparing two classification algorithms C4.5 and HT, respectively. C4.5 classification rules have a high accuracy which used

Table 2 Recovery ability—reciprocal—HOEFFDING TREE

Hoefding Tree	Q1	Q2	Q3	Recovery time
Ant	0.156	0.093601	0	0.0592803
Bat	0.093601	0.2652	0	0.0982802
Bee	0.0624	0.156	0.3588	**0.23868**
Cuckoo	0.1248	0.3588	0	0.
Elephant	0.1092	0.078001	0	0.0452403
Evolutionary	0.1404	0.4524	0	0.1638
Firefly	0.1092	0.078	0	0.04524
Flower	0.078001	0.0312	0	0.0249602
GA	0.156	0.1092	0	0.06396
Harmony	0.0624	0.0312	0.078001	0.0608405
PSO	0.1716	0.0468	0.093601	0.0951605
Wolf	0.1248	0.0312	0.093601	0.0811205
W1	0.2			
W2	0.3			
W3	0.5			

Table 3 Recovery ability—reciprocal—NAÏVE BAYS

Naïve Bays	Q1	Q2	Q3	Recovery time
Ant	0.1092	0.0624	0	0.04056
Bat	0.078	0.2184	0	0.08112
Bee	0.0312	0.0312	0	0.0156
Cuckoo	0.093601	0.2808	0	0.1029602
Elephant	0.093601	0.0624	0	0.0374402
Evolutionary	0.1092	0.2964	0	**0.11076**
Firefly	0.093601	0.0624	0	0.0374402
Flower	0.078001	0.0312	0	0.0249602
GA	0.1248	0.078001	0	0.0483603
Harmony	0.0468	0.0312	0.0624	0.04992
PSO	0.1248	0.0468	0.1092	0.0936
Wolf	0.1092	0.0312	0.078001	0.0702005
W1	0.2			
W2	0.3			
W3	0.5			

Table 4 Recovery ability—reciprocal—BOOSTADWIN

BOOSTADWIN	Q1	Q2	Q3	Recovery time
Ant	2.9328	4.0404	4.1652	3.88128
Bat	0	0	0	0
Bee	0	0	0	0
Cuckoo	0	0	0	0
Elephant	0	0	0	0
Evolutionary	1.7628	1.8876	4.5084	**3.17304**
Firefly	1.4664	1.9656	4.1808	2.97336
Flower	0	0	0	0
GA	1.2324	2.6676	2.73	2.41176
Harmony	0	0	0	0
PSO	0	0	0	0
Wolf	0	0	0	0
W1	0.2			
W2	0.3			
W3	0.5			

Table 5 Recovery ability—reciprocal—SGD

SGD	Q1	Q2	Q3	Recovery time
Ant	0.0156	0.0624	0.156	0.09984
Bat	0.0156	0.0312	0.093601	0.0592805
Bee	0.0156	0.0312	0.0156	0.0208
Cuckoo	0.0156	0.0624	0.156	0.09984
Elephant	0.0156	0.0624	0.1092	0.07644
Evolutionary	0.0156	0.0624	0.156	**0.09984**
Firefly	0.0312	0.0624	0.1248	0.08736
Flower	0.0156	0.0312	0.0624	0.04368
GA	0.1248	0.078	0	0.04836
Harmony	0.0468	0.0312	0.0624	0.04992
PSO	0.1404	0.0468	0.1092	0.09672
Wolf	0.093601	0.0312	0.078001	0.0670807
W1	0.2			
W2	0.3			
W3	0.5			

Table 6 Fluctuation Degree (accumulation accuracy)—HOEFFDING TREE

Hoeffding tree	Q1	Q2	Q3	Fluctuate degree
Ant	6	3	4	4.1
Bat	6	5	5	5.2
Bee	5	5	5	5
Cuckoo	6	5	6	5.7
Elephant	6	3	5	4.6
Evolutionary	8	5	4	5.1
Firefly	6	3	5	4.6
Flower	6	4	2	**3.4**
GA	7	3	4	4.3
Harmony	6	3	6	5.1
PSO	6	3	9	6.6
Wolf	7	3	7	5.8
W1	0.2			
W2	0.3			
W3	0.5			

Table 7 Fluctuation degree (accumulation accuracy)—NAÏVE BAYS

Naïve Bays	Q1	Q2	Q3	Fluctuate degree
Ant	6	3	4	4.1
Bat	6	5	5	5.2
Bee	9	3	2	3.7
Cuckoo	6	5	6	5.7
Elephant	6	3	5	4.6
Evolutionary	8	5	4	5.1
Firefly	6	3	5	4.6
Flower	6	4	2	**3.4**
GA	7	3	4	4.3
Harmony	6	3	6	5.1
PSO	6	3	9	6.6
Wolf	7	3	7	5.8
W1	0.2			
W2	0.3			
W3	0.5			

Table 8 Fluctuation degree (accumulation accuracy)—BOOSTADWIN

BOOSTADWIN	Q1	Q2	Q3	Fluctuate degree
Ant	7	7	10	8.5
Bat	5	11	10	9.3
Bee	5	10	7	7.5
Cuckoo	10	15	9	11
Elephant	8	7	6	6.7
Evolutionary	4	8	4	**5.2**
Firefly	6	11	7	8
Flower	11	11	7	8
GA	6	11	10	9.5
Harmony	7	9	12	10.1
PSO	6	8	5	6.1
Wolf	5	7	5	5.6
W1	0.2			
W2	0.3			
W3	0.5			

Table 9 Fluctuation degree (accumulation accuracy)—SGD

SGD	Q1	Q2	Q3	Fluctuate degree
Ant	6	8	5	6.1
Bat	6	8	5	6.1
Bee	6	6	4	5
Cuckoo	6	8	5	6.1
Elephant	7	6	6	6.2
Evolutionary	6	8	5	6.1
Firefly	6	8	5	6.1
GA	7	3	4	**4.3**
Harmony	6	3	6	5.1
PSO	6	3	9	6.6
Wolf	7	3	7	5.8
W1	0.2			
W2	0.3			
W3	0.5			

Table 10 Successive time—HOEFFDING TREE

Hoeffding Tree	Q1	Q2	Q3	Successive time
Ant	0.5616	0.5616	0.1092	0.3354
Bat	0.3432	0.3276	0.0312	0.18252
Bee	0.1716	0.156	0.0156	0.08892
Cuckoo	0.39	0.3744	0.0156	0.19812
Elephant	0.4056	0.39	0.0312	0.21372
Evolutionary	0.6396	0.624	0.1248	**0.37752**
Firefly	0.3588	0.3588	0.0468	0.2028
Flower	0.1404	0.1248	0.0156	0.07332
GA	0.4992	0.4836	0.0468	0.26832
Harmony	0.2028	0.2028	0.0312	0.117
PSO	0.3276	0.312	0.0312	0.17472
Wolf	0.234	0.234	0.0312	0.1326
W1	0.2			
W2	0.3			
W3	0.5			

Table 11 Successive time—NAÏVE BAYS

Naïve Bays	Q1	Q2	Q3	Successive time
Ant	0.3588	0.3588	0.0312	0.195
Bat	0.2496	0.234	0.0156	0.12792
Bee	0.1716	0.156	0.0156	0.08892
Cuckoo	0.3276	0.3276	0.0312	0.1794
Elephant	0.3432	0.3276	0.0156	0.17472
Evolutionary	0.3744	0.3744	0.0312	0.2028
Firefly	0.3276	0.312	0.0312	0.17472
Flower	0.156	0.1404	0.0312	0.08892
GA	0.421	0.4056	0.0312	**0.22152**
Harmony	0.1872	0.1872	0.0156	0.1014
PSO	0.2964	0.2652	0.0156	0.14664
Wolf	0.2184	0.2028	0.0156	0.11232
W1	0.2			
W2	0.3			
W3	0.5			

Table 12 Successive time—BOOSTADWIN

BoostAdwin	Q1	Q2	Q3	Successive time
Ant	0.1716	0.1248	0.093601	0.1185605
Bat	0.1248	0.078001	0.078001	0.0873608
Bee	0.1092	0.0468	0.0468	0.05928
Cuckoo	0.1404	0.1404	0.078001	0.1092005
Elephant	0.156	0.093601	0.093601	0.1060808
Evolutionary	0.1404	0.093601	0.093601	0.1029608
Firefly	0.1404	0.093601	0.093601	0.1029608
Flower	0.093601	0.093601	0.0468	0.0702005
GA	0.234	0.1872	0.1092	**0.15756**
Harmony	0.093601	0.078001	0.0468	0.0655205
PSO	0.1248	0.078001	0.078001	0.0873608
Wolf	0.1248	0.1092	0.078001	0.0967205
W1	0.2			
W2	0.3			
W3	0.5			

Table 13 Successive time—SGD

SGD	Q1	Q2	Q3	Successive time
Ant	0.1092	0.624	0.0156	0.04836
Bat	0.078001	0.0468	0.0156	0.0374402
Bee	0.0624	0.0312	0.0156	0.02964
Cuckoo	0.093601	0.0468	0.0156	0.0405602
Elephant	0.093601	0.0624	0.0156	0.0452402
Evolutionary	0.093601	0.0468	0.0156	0.0405602
Firefly	0.093601	0.0468	0.0156	0.0405602
Flower	0.0624	0.0468	0.0156	0.0405602
GA	0.4524	0.4368	0.0468	**0.24492**
Harmony	0.1872	0.1872	0.0156	0.1014
PSO	0.2964	0.2652	0.0312	0.15444
Wolf	0.2184	0.2028	0.0156	0.11232
W1	0.2			
W2	0.3			
W3	0.5			

to apply in cloud platform. HT is a popular choice of data stream mining algorithm which could be well used for fog computing. The simulation experiment is dedicated to IoT emergency services. Through collecting a large amount of data from gas sensor data to analyze all kinds of gas and then measure air quality. As a consequence of the experiment, C4.5 potentially gets high accuracy if the whole data are trained. But in the fog computing environment, the data are streaming in large amount nonstop into the data stream mining model. So, the model must be able to handle incremental learning from seeing only a portion of the data stream at a time. And it updates itself quickly each time fresh data is seen. Real-time latency and accuracy are required in IoT environment especially in fog environment; the experiment concludes that FS would have a slightly greater impact on C4.5. However, FS contribute, to ameliorate the performance of HT in fog environment. Moreover, Harmony search is an effective search method to strengthen the accuracy, time requirement and time cost for HT model in the data stream mining environment. Fog computing using HT coupled with FS-Harmony could have good accuracy, low latency and reasonable data scalability. In the second experiment, based on the close mean accuracy, there are three criteria. In the recovery time, the evolutionary algorithm is the best one. In the fluctuation degree, flower algorithm is the best one. In successive time, genetic algorithm is the best one. Through this tree criteria, we can compare algorithms and choose the best one.

Key Terminology and Definitions

Data Stream Mining Data stream mining is the process of extracting knowledge structures from continuous, rapid data records. A data stream is an ordered sequence of instances that in many applications of data stream mining can be read only once or a small number of times using limited computing and storage capabilities.

Swarm Search In computer science and mathematical optimization, a metaheuristic is a higher-level procedure or heuristic designed to find, generate, or select a heuristic (partial search algorithm) that may provide a sufficiently good solution to an optimization problem, especially with incomplete or imperfect information or limited computation capacity. Metaheuristics sample a set of solutions that is too large to be completely sampled. Metaheuristics may make few assumptions about the optimization problem being solved, and so they may be usable for a variety of problems.

Fog Computing Fog computing, also known as fog networking or fogging, is a decentralized computing infrastructure in which data, compute, storage and applications are distributed in the most logical, efficient place between the data source and the cloud. 'Fog computing essentially extends cloud computing and services to the edge of the network', bringing the advantages and power of the cloud closer to where data is created and acted upon.

Dr. Simon Fong graduated from La Trobe University, Australia, with a first Class Honours BEng. Computer Systems degree and a Ph.D. Computer Science degree in 1993 and 1998, respectively. Simon is now working as an Associate Professor at the Computer and Information Science Department of the University of Macau. He is a co-founder of the Data Analytics and Collaborative Computing Research Group in the Faculty of Science and Technology. Prior to his academic career, Simon took up various managerial and technical posts, such as systems engineer, IT consultant and e-commerce director in Australia and Asia. Dr. Fong has published over 432 international conference and peer-reviewed journal papers, mostly in the areas of data mining, data stream mining, big data analytics, metaheuristics optimization algorithms and their applications. He serves on the editorial boards of the Journal of Network and Computer Applications of Elsevier (I.F. 3.5), IEEE IT Professional Magazine, (I.F. 1.661) and various special issues of SCIE-indexed journals. Simon is also an active researcher with leading positions such as Vice-chair of IEEE Computational Intelligence Society (CIS) Task Force on 'Business Intelligence & Knowledge Management' and Vice-director of International Consortium for Optimization and Modelling in Science and Industry (iCOMSI).

Ms. Tengyue Li is currently an M.Sc. student major in E-Commerce Technology at the Department of Computer and Information Science, University of Macau, Macau SAR of China. She participated in the university the following activities: Advanced Individual in the School, Second Prize in the Smart City APP Design Competition of Macau and Top 10 in the China Banking Cup Million Venture Contest. Campus Ambassador of Love. Tengyue has internship experiences as aa Meituan Technology Company Product Manager from June to August 2017. She worked at Training Base of Huawei Technologies Co., Ltd. from September to October 2016. From February to June 2016, Tengyue worked at Beijing Yangguang Shengda Network Communications as data analyst. Lately, Tengyue involved in projects such as 'A Minutes' Unmanned Supermarket by the University of Macau Incubation Venture Project since September 2017.

Dr. Sabah Mohammed research interest is in intelligent systems that have to operate in large, nondeterministic, cooperative, survivable, adaptive or partially known domains. Although his research is inspired by his Ph.D. work back in 1981 (Brunel University, UK) on the employment of some Brain Activity Structures based techniques for decision making (planning and learning) that enable processes (e.g. agents, mobile objects) and collaborative processes to act intelligently in their environments to timely achieve the required goals. Dr. Mohammed is a full professor of Computer Science with Lakehead University, Ontario, Canada since 2001 and Adjunct Research Professor with the University of Western Ontario since 2009. He is the Editor-in-Chief of the international journal of Ubiquitous Multimedia (IJMUE) since 2005. Dr. Mohammed research touches many areas including Web Intelligence, Big Data, Health Informatics and Security of Cloud-Based EHRs among others.

Chapter 4
Pattern Mining Algorithms

Richard Millham, Israel Edem Agbehadji, and Hongji Yang

1 Introduction to Pattern Mining

In this chapter, we first look at patterns with their relevance of discovery to business. We then do a survey and evaluation, in terms of advantages and disadvantages, of different mining algorithms that are suited for both traditional and big data sources. These algorithms include those designed for both sequential and closed sequential pattern mining for both the sequential and parallel processing environments.

2 Pattern Mining Algorithm

Generally, data mining tasks are classified into two kinds: descriptive and predictive (Han and Kamber 2006). While descriptive mining tasks characterize properties of the data, predictive mining tasks perform inference on data to make predictions. In some cases, a user may have no information regarding what kind of patterns in data may be interesting and hence may like to search for different kinds of patterns.

A pattern could be defined as an event or grouping of events that occur in such a way that they deviate significantly from a trend and that they represent a significant

R. Millham (✉) · I. E. Agbehadji
ICT and Society Research Group, Durban University of Technology, Durban, South Africa
e-mail: richardm1@dut.ac.za

I. E. Agbehadji
e-mail: israeldel2006@gmail.com

H. Yang
Department of Informatics, University of Leicester, Leicester, England, UK
e-mail: Hongji.Yang@Leicester.ac.uk

S. J. Fong and R. C. Millham (eds.), *Bio-inspired Algorithms for Data Streaming and Visualization, Big Data Management, and Fog Computing*, Springer Tracts in Nature-Inspired Computing,
https://doi.org/10.1007/978-981-15-6695-0_4

difference from what would be expected of random variation (Iglesia and Reynolds 2005). In its simplest form, a pattern may illustrate a relationship between two variables (Hand et al. 2001) which may possess relevant and interesting information. Interesting can be denoted as implicit, previously unknown, non-trivial, and potentially useful information. The borders between models and patterns often intermix as models often contain patterns and other structures within data (Iglesia and Reynolds 2005).

An example of a pattern is a frequent pattern where a frequent pattern can have frequent itemsets, frequent subsequences, and frequent substructures. A frequent itemset represents a set of items that appear together in a dataset very often. A frequently occurring subsequence represents a pattern in which a user acquires an item first, followed by another item, and then series of itemset, is a (frequent) sequential pattern (Han and Kamber 2006). A frequent substructure represents different structural forms, namely graphs, trees, or lattices, which can be combined with itemsets or subsequences (Han and Kamber 2006). Thus, if substructures appear often, then it is referred to as frequent structured pattern.

There are various data mining algorithms that are used to reveal interesting patterns from both traditional data sources (such as relational databases) and big data sources. Big data sources provide additional challenges to these mining algorithms as the amount of data to be processed in very large and this data is both frequently generated and changed at high velocity. However, this data provides a venue to discover methods of mining interesting information that are relevant for businesses.

2.1 Sequential Pattern Mining

A sequential pattern (also known as a frequent sequence) is a sequence that has a support greater or equal to minimum threshold (Zhenxin and Jiaguo 2009). Sequential pattern mining, as per Aggarwal and Han (2014), is defined as association rule mining over a temporal database with stress being put on ordering of items. A sequential pattern algorithm has that property that every non-empty subsequence of a sequential pattern must occur frequently to illustrate the anti-monotonic (or downward closure) property of the algorithm (Aggarwal and Han 2014). In other words, a pattern that is denoted as frequent must have subsequences that are also frequent. Mining of a complete set of frequent subsequences that meet a minimum support threshold is identified as a shortcoming of sequential pattern mining (Raju and Varma 2015). This is because when long frequent sequences are mined, it can contain many frequent subsequences, which may create a huge number of frequent subsequences. Thus, the mining process becomes computationally expensive in terms of both time and memory space (Yan et al. 2003).

Algorithms for sequential pattern mining (Agrawal and Srikant 1995) include Apriori-based procedures, pattern-growth procedures, and vertical format-based procedures (Raju and Varma 2015). These algorithms are explained as follows.

2.1.1 Apriori-Based Method

The Apriori-based procedure is a level-wise method that is designed to generate frequent itemsets found within a dataset. The main principle of the Apriori method is that every subset of every frequent pattern is also frequent (which is also denoted as downward closure). These patterns are combined later through "Joins" (Aggarwal and Han 2014). These "Joins" facilitate the union of all patterns into a complete pattern. While the Apriori algorithm is being executed, a set of patterns is created as a candidate depiction of frequent patterns which is then tested. The purpose of this testing is to remove non-frequent patterns. This "candidate-generation-and-test" of Apriori leads to huge number of candidate, and consequently, more database scans is required to identify patterns (Tu and Koh 2010). The set of patterns are counted and pruned; thus, the Apriori-based method leads to high computational cost with respect to time and resources (Raju and Varma 2015). Thus, one of the major challenges with this policy is high computational cost. Additionally, when frequent itemsets are created as the output results, association rules with a confidence level that is greater than or equal to a minimum threshold can be generated (Kumar et al. 2007). One of the challenges with Apriori is that setting the minimum support threshold is largely based on intuition of user (Yin et al. 2013). Thus, if the threshold value is too low, many patterns can be generated which might require further filtering; however, if the threshold value is set high, it might produce no results (Yin et al. 2013). In order to address this problem of a huge number of results, pattern compression methods such as RPglobal and RPlocal (Han et al. 2007) have been applied; however, filtering these results required the use of computationally costly filtering algorithm. The use of the Apriori algorithm to mine patterns has the disadvantages of using "candidate-generation-and-test" with required user-set minimum support and confident thresholds, which may result in excessive computational costs.

The Apriori algorithm can be described in the following steps:

Step 1: explore all frequent itemsets
Step 2: obtain all frequent itemsets that are defined as itemsets whose items have an occurrence in the dataset greater than or equal to the minimum support threshold
Step 3: produce candidates from the newly obtained frequent itemsets
Step 4: prune the results to discover frequent itemsets
Step 5: discover association rules from frequent itemsets. The rules must meet both the minimum support threshold (as in Step 2) and the minimum confidence threshold value.

2.1.2 Pattern-Growth Methods

Pattern-growth method is based on a depth-first search. In the process finding a pattern, frequent pattern tree (FP tree) is created based on the idea of divide-and-conquer (Song and Rajasekaran 2006). Subsequently, this tree is separated into two and one part is chosen as the best branch. The chosen best branch is further developed by mining other frequent patterns. The frequent pattern-growth approach (Han et al. 2000) discovers frequent patterns without generating candidates of FP tree (Liu and Guan 2008). The advantages of FP tree are that it greatly compressed the result and produce much smaller dataset (Tu and Koh 2010). A second advantage of the FP tree is that it circumvents "candidate-generation-and-test" by "CONCATENATING" frequent items in a CONDITIONAL FP tree to ensure unnecessary candidate generation (Tu and Koh 2010). Pattern-growth-based algorithms include PrefixSpan (Pei et al. 2001) and FreeSpan (Han et al. 2000).

2.1.3 Vertical Format-Based Methods

The vertical format-based procedures use a vertical data structure to illustrate a sequence database (which are traditional database systems) (Raju and Varma 2015). The basis for using the vertical data structure is to facilitate quick computation and counting of the support threshold value on items. The quick computation emanates from the use of an id-list (such as binary) to link to a corresponding itemset. Vertical format-based algorithms include sequential pattern mining using bitmap representation (SPAM) (Ayres et al. 2002) and sequential pattern discovery (SPADE) (Zaki 2001).

The sequential pattern mining (SPAM) algorithm improves the support counting of items and on candidate generation from a dataset. The SPAM algorithm uses a "vertical bitmap depiction of the database" as part of its search strategy so as to improve candidate generation and support counting in very long sequential patterns (Ayres et al. 2002). Although this approach makes SPAM quicker in terms of computation than SPADE, SPAM consumes more memory space than SPADE (Raju and Varma 2015). An alternative representation of the vertical format is the horizontal format-based procedure that denotes a sequence database via a horizontal format with a sequence-id and a corresponding sequence of items. The disadvantage of this horizontal format is that it requires multiple scans over the data in order to produce a group of possible frequent sequences.

The sequential pattern discovery (SPADE) links a sequence database to a vertical data structure such that each item is taken as the center of observation that utilizes related sequence and event identifiers as datasets. SPADE decomposes original search space (i.e., in lattice form) into equivalent smaller parts (i.e., sub-lattices) what are then loaded and processed independently in main memory. While processing in main memory, each sub-lattice navigates the sequence tree in either breadth-first or depth-first methods and then uses a JOIN operation to concatenate two similar vertical id-list in its list structure. The disadvantage of SPADE is the high memory consumption requirement because each sub-lattice has to explore sequence of paths in a depth-first method, when a candidate is produced; it is stored in a lexicographic tree. Each sequence in the tree is either a sequence extended sequence (sequence produced by adding new transactions) or an itemset-extended sequence (sequence produced by appending an item to the last itemset). The disadvantages of SPADE are the high memory consumption and the usual challenges of candidate generation.

In order to improve candidate generation and the support counting of items, the SPAM algorithm uses a "vertical bitmap depiction" technique to improve on efficient candidate generation and support counting when sequential patterns are very lengthy (Ayres et al. 2002). Relatively, the SPAM algorithm is quicker than the SPADE algorithm because of the fast bitmap computation but at a high memory consumption cost than SPADE (Raju and Varma 2015). An alternative of the vertical format-based is a horizontal format-based method that embodies a sequence database using a horizontal format with sequence-id and a sequence of itemsets. The disadvantage of a horizontal format is that it requires multiple scans over the data to produce a set of potential frequent sequences.

In conclusion, sequential pattern mining entails subsequences with redundant patterns, which produces an exponential increase in patterns (Raju and Varma 2015) with consequent high computational cost.

2.1.4 Closed Sequential Pattern Mining Algorithms

Closed sequential pattern algorithm is an enhanced sequential pattern mining. This enhancement is in three ways. Firstly, closed sequential mining utilizes efficient use of search space pruning methods that greatly decrease the number of patterns produced (Huang et al. 2006). Secondly, closed sequential mining discovers more interesting patterns which decrease the encumbrance of the user being required to explore several patterns with the same minimum support threshold (Raju and Varma 2015). Thirdly, closed sequential pattern mining preserves every information in the entire pattern in a compact form (Cong et al. 2005). A closed sequential pattern is a frequent sequence which has no frequent super sequence (in other word, no larger itemset) with the same minimum support threshold value (in other words, the same occurrence frequency) (Yan et al. 2003). Consequently, close sequential pattern mining avoids finding patterns of super sequence with the same support threshold value (Huang et al. 2006). Some algorithms that are based on closed sequential pattern mining include ClaSP (Raju and Varma 2015), COBRA (Huang et al. 2006), CloSpan (Yan et al. 2003), and BIDE (Wang et al. 2007).

Clospan

The Clospan algorithm conducts data mining in two phases (Yan et al. 2003). The first phase produces closed sequential pattern as candidate set and keeps it in a prefix sequence lattice. The second phase does post-pruning to remove non-closed sequential patterns (Raju and Varma 2015). Conversely, this algorithm requires a very large search space for checking the closure of new patterns (Raju and Varma 2015).

Bidirectional Extension (BIDE)

The BIDE algorithm finds close patterns without maintaining candidate set. This is achieved by using depth-first search order to prune search space more deeply. It then performs closure check via a closure checking technique called "bidirectional extension." The "bidirectional extension" relates to "forward and backward directional extension." While backward direction extension prunes a search space, check for the closure of prefix patterns; forward directional extension is applied to construct prefix patterns and checks for closure of prefix patterns (Raju and Varma 2015). The backward directional extension stops the expansion of unnecessary patterns if the current prefix cannot be closed (Huang et al. 2006). In order to apply the BIDE algorithm, the BackScan approach within the algorithm first determines whether a prefix sequence can be removed; if not, it finds the number of "backward extension items," and then finds the number of "forward extension items"; and if there is "no backward extension item or forward extension item," then it indicates the "closed sequential patterns." When the BIDE algorithm is used, the benefit is that it "does not keep track of historical closed sequential patterns (or candidates)" for new patterns "closure checking" (Raju and Varma 2015; Huang et al. 2006). However, it requires multiple database scans which might consume computational time.

Closed Sequential Pattern Mining with Bi-phase Reduction Approach (COBRA)

The bi-phase reduction approach (COBRA) algorithm when applied to data mining, finds closed sequential pattern. The first reduction phase of this algorithm finds closed frequent itemsets and then encodes each mined item using a unique code (denoted as a C.F.I codes) in a new dataset. The second reduction phase then produces sequence extensions only from closed itemsets previously denoted using C.F.I code (which are closed itemsets). After reduction, mining is performed in three phases: (1) mining closed frequent itemsets (2) database encoding (3) mining closed sequential patterns. To enable more efficient pruning, a layer pruning approach was used to eliminate unnecessary enumeration (candidates) during the extensions of the same prefix pattern (Huang et al. 2006). This approach used two pruning methods: "Layer-Pruning and ExtPruning." On one hand, LayerPruning method helps to remove "non-closed branches," which avoid using more memory space in pattern checking. On the other hand, ExtPruning checks closure of pattern to remove "non-closed sequential patterns." This algorithm uses both vertical and horizontal database formats during search for patterns thereby decreasing the search time, which overcomes the disadvantages of pattern-growth method. The bi-phase reduction process both reduces the search spaces and duplicate combination but also has the advantage of avoiding the cost of matching item extensions (Huang et al. 2006). One advantage of COBRA is that it needs less memory space than the BIDE algorithm (Huang et al. 2006).

ClaSP

The ClaSP algorithm is used with data considered to denote a temporal dataset. Sequential patterns are produced using a vertical database format by the algorithm. The algorithm has two steps: the first step creates frequent closed candidates from the dataset which are then stored in memory; and the second step does recursive post-pruning to eliminate "all non-closed sequences" to obtain the final frequent closed sequences. The algorithm terminates when there are no non-closed sequences in candidates set of frequent items. In order to prune the search space, the ClaSP uses "CheckAvoidable" technique which outperforms the CloSpan (Raju and Varma 2015). Again, the ClaSP algorithm needs more main memory than other algorithms.

In the current dispensation of big data, these algorithms need to be enhanced for efficient discovery of patterns. The basis for the enhancement is due to the high communication cost of data transfer. The main issue is how can sequential pattern or "closed sequential pattern mining" algorithms be applied to big datasets to uncover hidden patterns with minimal computational cost and time given the characteristics (such as volume, velocity, etc.) associated with big data. Oweis et al. (2016) propose parallel data mining algorithm as a way to enhance data mining in big dataset.

2.2 *Parallel Data Mining Algorithm*

This algorithm helps to find pattern in a distributed environment. Ideally, searching for useful patterns in large volumes of data is very problematic (Cheng et al. 2013; Rajaraman and Ullman 2011). This because the user has to search through several uninteresting and not useful data which requires more computational time. The parallel data mining methods facilitate simultaneous computation to discover useful relationship among data items (Gebali 2011; Luna et al. 2011), thus reducing computational time while allowing large frequent pattern problems to be separated into smaller ones (Qiao et al. 2010). Examples of parallel sequential pattern mining algorithms are pSpade and HSPM (Shintani and Kitsuregawa 1998), and an example of "parallel closed sequential pattern mining" includes PAR-CSP (Cong et al. 2005).

2.2.1 Parallel Pattern Mining Algorithm

There are different forms of parallel sequential pattern mining algorithms which is parallel SPADE and the hash-based partition sequential pattern mining algorithm (HPSPM). While the HPSPM algorithm separates candidate sequences using a hash function, the parallel SPADE algorithm separates the search space into multiple suffix-based sets and processes task and data independently in parallel; after processing, the results are conglomerated into a group of frequent patterns (Cong et al. 2005). When data is separated into various partitions, it permits these partitions to independently compute the frequency count of patterns for efficient pruning of candidate sequences (Cong et al. 2005).

Besides HPSPM and parallel SPADE algorithms, this chapter explores the parallel Apriori algorithm and PARMA algorithm for association rule mining.

2.2.2 Parallel Apriori Algorithm

The parallel Apriori algorithm separates the candidate set into discrete subsets so as to adjust to the exponential growth of data which the traditional Apriori algorithm, which was outlined previously, splits candidate set into different subset to adapt to the exponential growth of data which the traditional Apriori algorithm which was earlier discussed, could not address due to the traditional Apriori algorithm generating an overfull group of candidate sets of frequent itemsets (Aggarwal and Rani 2013). Some big data mining frameworks (including the MapReduce framework) have utilized the parallel Apriori algorithm to attain quick synchronization as data size enlargens (Oweis et al. 2016). However, the cost of discovering output candidate sets remains high which caused Riondato et al. (2012) to propose a method of random sampling (separating) of a dataset into parallel separate sets and then filters and combines the output into a single set of results. The importance of this method [which is called the parallel randomized algorithm for approximate association rule mining (PARMA)] is that it decreases the computational cost of filtering and combining output results. This decrease increases the runtime performance of this PARMA algorithm in discovering association rules within large datasets (Oweis et al. 2016).

Algorithms which use randomization (Yang et al. 2015) in a big data environment use the following techniques such as randomized least squares regression, randomized k-means clustering, randomized kernel methods, randomized low-rank matrix approximation, and randomized classification (regression). The benefit of using a randomized algorithm is that the discovery method is quicker and robust; it exploits the advantages of parallel algorithms and, consequently, it is efficient. Although randomization has the advantage in that is quicker and reduced data size, it also has the disadvantage of being prone to error when discovering the properties of data attributes (Yang et al. 2015). Optimization techniques, as specified by Yang et al. (2015), affords an optimal solution, improves the convergence rate of data, and discovers properties of functions yet these techniques are vulnerable to high computational and communication cost. Blending the advantages of randomization and optimization leads to an efficient search algorithm and a decent initial solution (Yang et al. 2015). It is also possible to associate these properties of data attributes to the frequently changed or frequently used aspect of data. Consequently, discovering the frequently changed or frequently used attributes through randomization and optimization may produce interesting patterns which the present traditional techniques of data mining (such as sequential and extension of sequential pattern mining algorithms) do not address.

2.2.3 Parallel Closed Sequential Pattern Mining (Par-CSP) Algorithms

This algorithm facilitates mining on different systems based on the principle of divide-and-conquer to partition their task with the consequence of decreased overhead communication cost (Cong et al. 2005). This algorithm uses the BIDE algorithm to mine closed sequential patterns without keeping the generated candidate dataset.

3 Conclusion

In this chapter, we examined various traditional data mining algorithms including those of types parallel sequential pattern mining and "parallel closed sequential pattern mining," "closed sequential pattern mining," and sequential pattern mining. These algorithms possess challenges and possible solutions with respect to candidate generation, pattern pruning, and the setting of user thresholds to filter potentially interesting patterns. A tabular summary of these algorithms, which include their approach, advantages, and limitations, is outlined in Appendix. These algorithms often focus on frequent itemset patterns, which are patterns that are often of interest to business due to their regularity.

Key Terminology and Definitions

Pattern could be defined as an event or grouping of events that occur in such a way that they deviate significantly from a trend and that they represent a significant difference from what would be expected of random variation.

Data mining is the application of an algorithm to a dataset to extract patterns or to construct a model to represent a higher level of knowledge about the data.

Appendix: Summary on Mining Algorithms

Author	Algorithm	Mining approach	Approach used	Advantages	Limitations
Aggarwal and Han (2014)		Apriori-based methods			However, a candidate-generation-and-test strategy produces a large number of candidate sequences and also requires more database scan (Tu and Koh 2010) when there are long patterns
Han et al. (2000)		Pattern-growth methods		Without candidate generation compressed database structure which is smaller than original dataset	
Han et al. (2000)	FreeSpan	Pattern-growth methods	sequential patterns by partitioning		
Pei et al. (2001)	PrefixSpan	Pattern-growth methods	Pseudo-projection technique for constructing projected databases	Does not generate-and-test any candidate sequence that do not exist in a projected database	Projected database requires more storage space, extra time is required to scan the projected database
		Sequential pattern mining	Vertical format-based methods	Fast computation of support counting	
Zaki (2001)	SPADE	Sequential pattern mining	Vertical format-based methods Either breadth-first or depth-first manner		Consumes more memory

(continued)

(continued)

Author	Algorithm	Mining approach	Approach used	Advantages	Limitations
Ayres et al. (2002)	SPAM	Sequential pattern mining	Vertical format-based methods Traverses the sequence tree in a depth-first manner	A vertical bitmap of the database	Consumes more memory space
Agrawal and Srikant (1995)	AprioriAll	Sequential pattern mining	Horizontal format-based method		Consumes more memory space
		Closed sequential pattern mining		Efficient use of search space pruning, reduced number of pattern, find more interesting patterns	
Yan et al. (2003)	CloSpan	Closed sequential pattern mining	Prefix sequence lattice, post-pruning		Huge search space for checking the closure of new patterns
Wang et al. (2007)	BIDE	Closed sequential pattern mining	Depth-first search order. perform closure checking (bidirectional Extension)	Without candidate maintenance, does not keep track of historical closed sequential patterns	Multiple database scan, more computational time
Huang et al. (2006)	COBRA	Closed sequential pattern mining	Bi-phase Reduction Approach, item encoding, pruning methods (LayerPruning and ExtPruning), vertical and horizontal database formats	Reduce searching space	Requires large memory space
Raju and Varma (2015	ClaSP	Closed sequential pattern mining	Vertical database format, Frequent Closed Candidates, recursive post-pruning (CheckAvoidable for pruning the search)		Requires more main memory

(continued)

(continued)

Author	Algorithm	Mining approach	Approach used	Advantages	Limitations
Han et al. (2002)		Top-K closed sequential pattern mining	Descending order of support	Without specifying minimum support	Users must decide the value of *k*, prior knowledge of database required
Hirate et al. (2004)	TF^2P-growth	Top-K closed sequential pattern mining	Descending order of support	Does not require the user to set any threshold value *k*, output of frequent patterns to user sequentially and in chunks	Time-consuming to be checking all chunk size
Wang et al. (2014)	BI-TSP	Top-K closed sequential pattern mining	bidirectional checking scheme, minimum length constraint, dynamically increase support of *k*		
Raju and Varma (2015)	CSpan	Top-K closed sequential pattern mining	Depth-first search, occurrence checking method for early detection of closed sequential patterns, constructs the projected database		Projected database requires more storage space, extra time is required to scan the projected database

References

Aggarwal, C. C., & Han, J. (2014). *Frequent pattern mining*. Springer International Publishing Switzerland. Available https://doi.org/10.1007/978-3-319-07821-2_3.

Aggarwal, S., & Rani, B. (2013). Optimization of association rule mining process using Apriori and ant colony optimization algorithm.

Agrawal, R., & Srikant, R. (1995). Mining sequential patterns. In: *Proceedings of International Conference Data Engineering (ICDE '95)* (pp. 3–14).

Ayres, J., Gehrke, J., Yiu, T., & Flannick, J. (2002). Sequential pattern mining using a bitmap representation. In: *Proceedings of ACM SIGKDD Int'l Conf. Knowledge Discovery and Data Mining (SIGKDD' 02)* (pp. 429–435).

Cheng, S., Shi, Y., Qin, Q., & Bai, R. (2013). *Swarm intelligence in big data analytics*. Berlin: Springer.

Cong, S., Han, J., & Padua, D. (2005). *Parallel mining of closed sequential patterns*. Available http://hanj.cs.illinois.edu/pdf/kdd05_parseq.pdf.

Gebali, F. (2011). *Algorithms and parallel computing*. Hoboken, NJ: Wiley.

Han, J., & Kamber, M. (2006). *Data mining concepts and techniques*. Morgan Kaufmann.

Han, J., Cheng, H., Xin, D. & Yan, X. (2007). Frequent pattern mining: current status and future directions. *Data mining and knowledge discovery, 15*(1), 55–86.

Han, J., Pei, J., Mortazavi-Asl, B., Chen, Q., Dayal, U., & Hsu, M. C. (2000). FreeSpan: Frequent pattern projected sequential pattern mining. In *Proceedings of ACM SIGKDD International Conference on Knowledge Discovery and Data Mining (SIGKDD'00)* (pp. 355–359).

Han, J., Wang, J., Lu, Y., & Tzvetkov, P. (2002). Mining top-k frequent closed patterns without minimum support. In *Proceedings of IEEE ICDM Conference on Data Mining.*

Hand, D., Mannila, H., & Smyth, P. (2001). *Principles of data mining*. London: The MIT Press.

Hirate, Y., Iwahashi, E., & Yamana, H. (2004). TF2P-growth: An efficient algorithm for mining frequent patterns without any thresholds. http://elvex.ugr.es/icdm2004/pdf/hirate.pdf.

Huang, K., Chang, C., Tung, J., & Ho, C. (2006). COBRA: Closed sequential pattern mining using bi-phase reduction approach.

Iglesia, B., & Reynolds, A. (2005). The use of meta-heuristic algorithms for data mining.

Kumar, V., Xindong, W., Quinlan, J. R., Ghosh, J., Yang, Q., Motoda, H., et al. (2007). *Top 10 algorithms in data mining*. London: Springer.

Liu, Y., & Guan, Y. (2008). Fp-growth algorithm for application in research of market basket analysis. In *2008 IEEE International Conference on Computational Cybernetics* (pp. 269–272). IEEE.

Luna, J. M., Romero, R. J., & Ventura, S. (2011). *Design and behavior study of a grammar-guided genetic programming algorithm for mining association rules*. London: Springer.

Oweis, N. E., Fouad, M. M, Oweis, S. R., Owais, S. S., & Snasel, V. (2016). A novel mapreduce lift association rule mining algorithm (MRLAR) for big data. *International Journal of Advanced Computer Science and Applications (IJACSA), 7*(3).

Pei, J., Han, J., Mortazavi-Asl, B., Wang, J., Pinto, H., Chen, Q., et al. (2001). PrefixSpan: Mining sequential patterns efficiently by prefix-projected pattern growth. In *Proceedings of the International Conference on Data Engineering (ICDE)* (pp. 215–224).

Qiao, S., Li, T., Peng, J., & Qiu, J. (2010). Parallel sequential pattern mining of massive trajectory data.

Rajaraman, A., & Ullman, J. (2011). *Mining of massive datasets*. Cambridge University Press.

Raju, V. P., & Varma, G. P. S. (2015). Mining closed sequential patterns in large sequence databases. *International Journal of Database Management Systems (IJDMS), 7*(1).

Riondato, M., DeBrabant, J. A., Fonseca, R., & Upfal, E. (2012). PARMA: A parallel randomized algorithm for approximate association rules mining in MapReduce. In *Proceedings of the 21st ACM International Conference on Information and Knowledge Management* (pp. 85–94). ACM.

Shintani, T., & Kitsuregawa, M. (1998). Mining algorithms for sequential patterns in parallel: Hash based approach. In *Proceedings of Pacific-Asia Conference on Research and Development in Knowledge Discovery and Data Mining* (pp. 283–294).
Song, M., & Rajasekaran, S. (2006). A transaction mapping algorithm for frequent itemsets mining. *IEEE Transactions on Knowledge and Data Engineering, 18*(4).
Tu, V., & Koh, I. (2010). A tree-based approach for efficiently mining approximate frequent itemset.
Wang, J., Han, J., & Li, C. (2007). Frequent closed sequence mining without candidate maintenance. *IEEE Transactions on Knowledge and Data Engineering, 19*(8), 1042–1056.
Wang, J., Zhang, L., Liu, G., Liu, Q., & Chen, E. (2014). On top-k closed sequential patterns mining. In *11th International Conference on Fuzzy Systems and Knowledge Discovery.*
Yan, X., Han, J., & Afshar, R. (2003). CloSpan: Mining closed sequential patterns in large databases. In: *Proceedings of SIAM International Conference on Data Mining (SDM '03)* (pp. 166–177).
Yang, T., Lin, Q., & Jin, R. (2015). Big data analytics: Optimization and randomization. In *Proceedings of the 21th ACM SIGKDD International Conference on Knowledge Discovery and Data Mining* (pp. 2327–2327). Available on https://homepage.cs.uiowa.edu/~tyng/kdd15-tutorial.pdf.
Yin, J., Zheng, Z., Cao, L., Song, Y., & Wei, W. (2013). Efficiently mining top-k high utility sequential patterns. In *Proceedings of 2013 IEEE 13th International Conference on Data Mining* (pp. 1259–1264).
Zaki, M. J. (2001). Parallel sequence mining on shared-memory machines. *Journal of Parallel and Distribution Computing, 61*(3), 401–426.
Zhenxin, Z., & Jiaguo, L. (2009). Closed sequential pattern mining algorithm based positional data. In *Advanced Technology in Teaching—Proceedings of the 3rd International Conference on Teaching and Computational Science (WTCS)* (pp. 45–53).

Richard Millham is currently an associate professor at the Durban University of Technology in Durban, South Africa. After thirteen years of industrial experience, he switched to the academe and has worked at universities in Ghana, South Sudan, Scotland, and the Bahamas. His research interests include software evolution, aspects of cloud computing with m-interaction, big data, data streaming, fog and edge analytics, and aspects of the Internet of things. He is a Chartered Engineer (UK), a Chartered Engineer Assessor, and Senior Member of IEEE.

Israel Edem Agbehadji graduated from the Catholic University College of Ghana with B.Sc. Computer Science in 2007, M.Sc. Industrial Mathematics from the Kwame Nkrumah University of Science and Technology in 2011 and Ph.D. Information Technology from Durban University of Technology (DUT), South Africa, in 2019. He is a member of ICT Society of DUT Research group in the Faculty of Accounting and Informatics; and *IEEE* member. He lectured undergraduate courses in both DUT, South Africa, and a private university, Ghana. Also, he supervised several undergraduate research projects. Prior to his academic career, he took up various managerial positions as the management information systems manager for the National Health Insurance Scheme; the postgraduate degree program manager in a private university in Ghana. Currently, he works as a Postdoctoral Research Fellow, DUT, South Africa, on joint collaboration research project between South Africa and South Korea. His research interests include big data analytics, Internet of things (IoT), fog computing, and optimization algorithms.

Hongji Yang graduated with a Ph.D. in Software Engineering from Durham University, England with his M.Sc. and B.Sc. in Computer Science completed at Jilin University in China. With over 400 publications, he is full professor at the University of Leicester in England. Prof Yang has been an IEEE Computer Society Golden Core Member since 2010, an EPSRC peer review college member since 2003, and Editor in Chief of the International Journal of Creative Computing.

Chapter 5
Extracting Association Rules: Meta-Heuristic and Closeness Preference Approach

Richard Millham, Israel Edem Agbehadji, and Hongji Yang

1 Introduction to Data Mining

Data mining is an approach used to find hidden and complex relationships present in data (Sumathi and Sivanandam 2006) with the objective to extract comprehensible, useful and non-trivial knowledge from large data sets (Olmo et al. 2011). Although there are many hidden relationships in data to be discovered, this chapter, we focus on association rule relationships which are explored using association rule mining.

Data mining algorithms find hidden and complex relationships present in data (Sumathi and Sivanandam 2006). Often, existing data mining algorithms are focused on the frequency of items without considering other dimensions that commonly occur with frequent data such as time. Basically, the frequency of item is computed by counting the occurrence in each transaction (Song and Rajasekaran 2006) to find interesting patterns. Usually, in frequent itemset mining, an itemset is regarded as interesting if its occurrence exceeds a user-specified threshold (Fung et al. 2012; Han et al. 2007) in terms of minimum support threshold. For example, when a set of items in this case printer and scanner appears frequently together, then it is said to be frequent itemset. When an item satisfies a set of parameters that is set as minimum support threshold, then it is considered as having an interesting pattern. However, this interesting pattern requires a user to take an action but when time for the action

R. Millham (✉) · I. E. Agbehadji
ICT and Society Research Group, Durban University of Technology, Durban, South Africa
e-mail: richardm1@dut.ac.za

I. E. Agbehadji
e-mail: israeldel2006@gmail.com

H. Yang
Department of Informatics, University of Leicester, Leicester, England, UK
e-mail: Hongji.Yang@Leicester.ac.uk

S. J. Fong and R. C. Millham (eds.), *Bio-inspired Algorithms for Data Streaming and Visualization, Big Data Management, and Fog Computing*, Springer Tracts in Nature-Inspired Computing,
https://doi.org/10.1007/978-981-15-6695-0_5

is not indicated, it poses a challenge particularly when data set is characterized with velocity (that is, data to be processed quickly). Therefore, the use of frequency to measure pattern interestingness (Tseng et al. 2006) when selecting actionable sequence may be expanded to cover the time dimension. For instance, a frequent item that considers a time dimension is when a customer who bought a printer, buys a scanner after one week, and then buys a CD after another one week. Thus, an example of a sequential rule is a customer, after buying a printer which is the antecedent, then buys a scanner in one week afterwards which represents the consequent. Thus, this sequential rule has a component of time-closeness dimension of one month between items. The time dimension enables the disclosure of interesting patterns within the selected time interval. Similarly, the numeric dimension in this instance is that, a customer who bought a printer at a price buys a scanner at a price, and then buys a CD at a price. Thus, the numeric dimension might present an interesting pattern as time changes. Therefore, considering both time and numeric dimension of frequent items is significant when mining association rules from frequent items. Han et al. (2007) indicated that frequent patterns are itemsets, subsequences or substructures that appear in a data set with frequency not less than a user-specified threshold.

In this chapter, we discuss how to extract interval rule with time intervals. Intuitively, an interval pattern is a vector of intervals, where each dimensions corresponding to a range of values of a given attribute.

2 Meta-heuristic Search Methods to Data Mining of Association Rule

Wu et al. (2016), define big data analytics (or big data mining) as the discovery of actionable knowledge patterns from quality data. Quality is defined in terms of accuracy, completeness and consistency (Garcia et al. 2015) of patterns. In other words, big data analytics is data mining process which involves the extraction of useful information (that is information that help organizations make more informed business decision, etc.) from very large volumes of data set (Cheng et al. 2013; Rajaraman and Ullman 2011).

Meta-heuristic algorithms play a significant role in extracting useful information from a big data analytics framework. Conceptually, a meta-heuristic is used to define heuristic methods applicable to a set of widely different problems (Dorigo et al. 2006) in order to find the best possible solution to a problem. Although problems (relating to finding optimal results) may require different approach, the meta-heuristic methods provide a general-purpose algorithmic framework applicable to different problems with relatively few modifications (Dorigo et al. 2006). Meta-heuristic methods are based on successful characteristics of species of a particular animal. Example of the characteristics includes how animals search for its prey in a habitat and includes swarm movement patterns of animals and insects found in nature (Tang et al. 2012). The algorithms that are developed from animals' behaviour are often referred to as bio-inspired algorithms. Meta-heuristic algorithms are used as search methods to solve different problems. The search methods are both in-breadth and in-depth searches to enable adequate exploration and exploitation of the search space. The meta-heuristic algorithms are able to find best optimal solution within a given optimization problem domain such as the discovery of association rules.

In many cases, data mining algorithms generate extremely large number of complex association rules (that is, of the form $V_1V_2...V_{n-1} \rightarrow V_n$, where V represents set of rules, n is the number of items being considered) that limit the usefulness of data mining results (Palshikar et al. 2007); thus, the proper selection of interesting rules is very important in reducing this number.

The following are meta-heuristic algorithms that can be applied to data mining: Genetic algorithm (GA) (Darwin 1868 as cited by Agbehadji 2011), Particle swarm optimization algorithm (PSO), ant colony optimization (ACO) (Dorigo et al. 1996) and wolf search algorithm (WSA) (Tang et al. 2012).

2.1 Genetic Algorithm

In applying genetic algorithm which has its basis in theory of "natural selection" (Darwin 1868 as cited by Agbehadji 2011) enables species considered as weak and cannot adapt to the conditions of the habitat are eliminated while species considered as strong and can adapt to the habitat survive. Thus, natural selection is based on the notion that strong species have greater chance to pass their genes to future generations, while weaker species are eliminated by natural selection. Sometimes, there are random changes that occur in genes due to changes within external environments of species, which will cause new future species that are produced to inherit different genetic characteristics. At the stage of producing new species, individuals are selected, at random, from the current population within the habitat to be parents and use them to produce the children for the next generation, thus successive generations are able to adapt to the habitat in respect of time. Terminology used in genetic algorithm to represent population member is string or chromosomes. These chromosomes are made of discrete units called genes (Railean et al. 2013) which are binary representation such as 0 and 1. There are rules to govern the combination of parents to form children. These rules are referred to as operator, namely the crossover, mutation and selection methods. The notion of crossover consists of interchanging solution values of particular variables; while mutations consist of random value changes to a single parent. The children produced by the mating of parents are tested and only children that pass the test (that is, the survival test) are then chosen as parents for the next generation. The survival test acts as a filter for selecting the best species. The adaptive search process of genetic algorithm has been applied to solve problems of association rule mining without setting of minimum support and minimum confidence value. Qodmanan et al.'s (2011) approach is multistage that first finds frequent itemset and then extracts association rules from frequent itemsets.

The approach combines the frequent pattern (FP) tree algorithm and genetic algorithm to form a multiobjective fitness function with support, confidence thresholds and be able to obtain interesting rules. This approach enables a user to change the fitness function so that the order of items is considered on importance of rules.

2.2 *Swarm Behaviour*

Refer to the previous chapter (chapter one) on the behaviour of particle swarm (Kennedy and Eberhart 1995; Krause et al. 2013). Among the particle swarm algorithms are the Firefly (Yang 2008), Particle swarm (Kennedy and Eberhart 1995), Bats (Yang 2009), etc.

2.2.1 Particle Swarm Optimization

Kuo et al. (2011) applied PSO on stock market database to measure investment behaviour and stock category purchasing. The method first searches for the optimum fitness value of each particle and then finds its corresponding support and confidence as minimal threshold values after the data are transformed into binary data type with each stored as either 0 or 1. When binary data type is used, it reduces the computational time required to scan the entire data sets in order to find the minimum support value without the user's intuition. The significance of this approach is that it helps with the automatic determination of support value from the data set thereby improving on the quality of association rules and computational efficiency (Kuo et al. 2011) as the search space enables tuning of several thresholds so as to select the best threshold. This saves a user the task of finding the best threshold by intuition.

Sarath and Ravi (2013) formulated a discrete/combinatorial global optimization approach that uses a binary PSO to mine association rules without specifying the minimum support and minimum confidence of items unlike the Apriori algorithm. The fitness function is used to evaluate the quality of the rules expressed as the product between the support and the confidence. The fitness function ensures the support and confidence are binary between 0 and 1. The proposed binary PSO algorithm consists of two parts, the preprocessing and the mining. The pre-processing part calculates the fitness values of the particle swarm in order to transform the data into binary data type to avoid computational time complexity; the mining part of the algorithm uses the PSO algorithm to mine association rules. Sarath and Ravi (2013) indicated that binary PSO can be used as an alternative to the Apriori algorithm and the FP-growth algorithm as it allows the selection of rules that satisfies the minimum support threshold.

2.2.2 Ant Colony Optimization

The ant colony optimization (ACO) (Dorigo et al. 1996) is a method that is based on foraging behaviour of real ants in their search for the shortest paths to food sources in their natural environment. When a source of food is found, ants deposit pheromone to mark their path for other ants to traverse. Pheromone is an odorous substance which is used as a medium of indirect communication between ants. The quantity of pheromone depends on the distance, quantity and quality of food source (Al-Ani 2007). However, the pheromone substance which decays or evaporates with time prevents ants from converging thereby ants can explore other sources of pheromone substances within its habitat (Stützle and Dorigo 2002). In a situation where an ant is lost, it moves at random in search for a laid pheromone, likely ants will follow the path that reinforces the pheromone trails. Thus, ants make probabilistic decisions on updating their pheromone trail and local heuristic information (Al-Ani 2007) to explore larger search areas. The ACO has been applied to solve many optimization-related problems, including data mining problems, where it was shown to be efficient in finding best possible solutions. In data mining, frequent itemset discovery is an important factor in implementing association rule mining. Kuo and Shih (2007) proposed a model that uses the ant colony system to first find best global pheromone and secondly generates association rules after a user specifies more than one attribute and defines two or more search constraints on an attribute. Constraint-based mining enables users to extract rules of interest to their needs and the consequent computational speed was faster, thus improving the efficiency of mining tasks. Kuo and Shih (2007) indicated that the constrained-based mining provided condensed rules contrary to those used by the Apriori method. Additionally, the computational time was reduced since the database was scanned only once to disclose the mined association results. The use of constraint conditions reduces search time during mining stage; however, the challenge with these constraints is finding a feasible method that can merge many similar rules generated in the mining results.

2.2.3 Wolf Search Algorithm

In this subsection, we refer the reader to wolf search algorithm (WSA) which is already discussed in previous chapter (chapter one) (Tang et al. 2012; Agbehadji et al. 2016).

2.2.4 Bat Algorithm

In this subsection, we refer the reader to bat algorithm (Yang 2010; Fister et al. 2014) which was explained in previous chapters (chapter one) of this book. Moreover, variant of bat algorithm includes sampling, improved bat algorithm (SIBA) (Wei et al. 2015). The SIBA was implemented on the cloud model to search for a candidate of frequent itemsets from a large data set according to a sample size of data. The basis of SIBA was to reduce the computational cost of scanning frequent itemsets. In order to achieve this, the approach used a fixed length of frequent itemset to mine top-k frequent l item set. The fixed iteration steps and the fixed population size of SIBA reduce the computational time (Wei et al. 2015). Wei et al. (2015) and Heraguemi et al. (2015) indicated that bat algorithm performs faster than Apriori and FP-growth and it was also robust than PSO and GA (Wei et al. 2015).

Although many meta-heuristic algorithms (such as Bat) use fixed parameters by using pre-tuned algorithm-dependent parameters, the parameters are controlled by the bio-inspired behaviour to vary the value of parameter, at each iteration process (Wei et al. 2015).

3 Data Mining Model

The data mining model describes various stages of a data mining process. These stages are important in determining the interestingness of rules at some stages in the data mining model as illustrated in Fig. 1.

Figure 1 represents a data mining model shows the stages where interestingness measures are applied during the pre-processing and post-processing stages. Initially, raw data is loaded into the model to yield interesting patterns such as association rules as output (Geng and Hamilton 2006).

During the pre-processing stage, pre-processing measures are used to prune uninteresting patterns/association rules in the mining stage so as to reduce the search space and to improve on mining efficiency. Measures that are applied at the pre-processing stage to determine the interestingness of rules adhere to the "anti-monotone property" states that the value assigned to a pattern must not be no greater than its sub-patterns (Agrawal and Srikant 1994a). An example of pre-processing measure includes the support measure, support-closeness preferences (CP) measure and F-measure. The difference among these measures is that the support measure uses the minimum support threshold, where a user may set a threshold to eliminate information that does not appear enough times in a database; the support-CP-based measure (Railean et al. 2013) selects patterns with closer antecedent and consequent so as to fulfil the anti-monotone property (principle of Apriori); and F-measure is a statistical approach

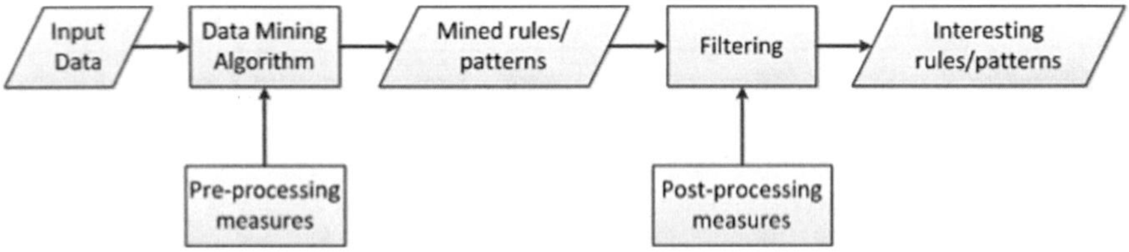

Fig. 1 Interestingness measure for data mining model. *Source* Railean et al. (2013)

that is based on precision (i.e. it gives more true positives) and recall (i.e. it gives more False Negatives)

During the post-processing stage, post-processing measure is used to filter the extracted information and obtain final pattern/rules in a search space. An example of post-processing measure is the use of confidence measure and lift measure (Mannila et al. 1997). Lift measure is defined as the ratio of the actual support to the expected support if support of one item (X) and another item (Y) was independent (Jagtap et al. 2012). Railean et al. (2013) proposed the following post-processing measures such as Closeness Preference (CP), Modified CP and Modified CP Support-Confidence (MCPsc) based measure and Actionability (Silberschatz and Tuzhilin 1995).

Closeness Preference (CP) takes into consideration a time interval to meet the user's preference to select rules with closer antecedent and consequent; Modified CP (MCP) measure is used to extract and rank patterns; the Modified CP Support-Confidence (MCPsc)-based measure selects patterns with closer itemsets and this measure does not fulfil the anti-monotone property (Railean et al. 2013).

The interestingness measures proposed by Railean et al. (2013) have been applied on various real data sets to show patterns and rules in Web analysis (for predicting the pages that will be visited), marketing (to find the next items that would be bought) and network security (to prevent intrusion from unwanted packages). The results validated the use of interestingness measure proposed by Railean et al. (2013) on real-world events.

Actionability (Silberschatz and Tuzhilin 1995) is a post-processing measure that determines whether or not a pattern is interesting by filtering out redundant patterns (Yang et al. 2003) and by disclosing if a user can get benefits/value [e.g. profit (Wang et al. 2002)] from taking actions based on these patterns. Cao et al. (2007) indicated that actionability of a discovered pattern must be assessed in terms of domain user needs. The needs of a user may be diverse but specifying the needs in terms time and numeric dimension is important.

Geng and Hamilton (2006) indicated that determination of interestingness measure, at both the pre-processing and post-processing stages, should be performed in three steps that are grouping of patterns, determining of preference and ranking of patterns. First, group each pattern as either interesting or uninteresting. Secondly, determine a user preferences in terms of a pattern that is considered as interesting compared to another pattern. Thirdly, ranking of the preferred patterns. These steps provide a framework in determining any interesting measure. As part of these steps, a pattern should adhere to defined properties so as to avoid ambiguity and create a general notion of interestingness measure. Piatetsky-Shapiro (1991) suggested properties that rules, which form a pattern, must adhere to in order for it to be considered as an interesting rule:

Property 1 "*An interesting measure is* 0 *meaning if* A *and* B *are statistically independent i.e. when* $P(AB) = P(A) \cdot P(B)$, *where P represents the probability and both antecedent* (A) *and the consequent* (B) *of the rule are statistically independent*" (Piatetsky-Shapiro 1991).

Property 2 "*An interestingness measure should monotonically increase with P(AB) when P(A) and P(B) remain the same. Thus, the higher confidence value the more interesting the rule is*" (Piatetsky-Shapiro 1991).

Property 3 "*An interestingness measure should monotonically decrease with P(A) (or P(B)) when P(AB)* and *P(B) (or P(A)) remain the same. This implies, when P(A) (or P(B)) and P(AB) are the same or has not changed, the rule interestingness monotonically decreases with P(B) thus the less interesting the rule is*" (Piatetsky-Shapiro 1991).

In this subsection, we look at the time closeness of items; therefore, the time-closeness preferences (CP) models are given prominence since traditional support and confidence used for extracting association rules do not consider the time closeness of rules. When time closeness is defined, it helps the user to identify items upon which to take an action. For instance, when speed in respect to time is significant in finding interesting patterns in a large set of data, the time-closeness preference model can show the time difference between items over an entire sequence of items (Railean et al. 2013). The smaller the time differences, the closer the items. The key question is how does the CP model relates to the steps provided by Geng and Hamilton (2006). The CP model allows the user to define time closeness of items and rank discovered patterns taking into consideration the time dimension. The aim of CP interestingness measure is to select the "strong" rules that represent frequency of antecedent and consequent of items, and with respect to closeness between itemsets A and B of the rule where the consequent B is as close as possible to the antecedent A in most of the cases with respect to time. Additionally, CP interestingness measure can be used to rank rules based on user's preferences (Railean et al. 2013). The advantage of CP model is the efficient extraction of rules and ranking when time closeness is of importance. It may be desirable to rank rules at the same level if the time-difference between the itemsets is no greater than a certain time denoted by σ_t, and then, to decrease the importance with respect to time by imposing a time-window ω_t (Railean et al. 2013).

The notion of time closeness, as an interestingness measure, can be defined (Railean et al. 2013) as follows:

Definition 1 Time Closeness—sequential rules. Let ω_t be a time interval and W a time window of size ω_t. An itemsets A and B with time-stamps t_A and t_B, respectively, are ω_t-close iff $|t_B - t_A| \leq \omega_t$. When considering a sequential rule $A \rightarrow B$, i.e. $t_B \geq t_A$, A and B are ω_t-close iff $t_B - t_A \leq \omega_t$.

Definition 2 Closeness Measure—sequential rules. Let σ_t be a user-preferred time interval, $\sigma_t < \omega_t$. A closeness measure for a ω_t close rule $A \rightarrow B$ is defined as a decreasing function of $t_B - t_A$ and $1/\sigma_t$ such that if $t_B - t_A \leq \sigma_t$ then the measure should decrease slowly while if $t_B - t_A > \sigma_t$ then the measure should decrease rapidly.

Definitions 1 and 2 take into consideration a single user preference in defining a rule. However, it is possible to have two user preferences in a time interval. Thus, a third definition was stated to define a pattern when two user preferences are required. The third definition (Railean et al. 2013) is stated as follows:

Definition 3 Time-Closeness Weight—sequential patterns. Let σ_t and ω_t be two user-preference time intervals, subject to. $\sigma_t < \omega_t$, and ω_t being the time after which the value of the weight passes below 50%. The time-closeness weight for a pattern $P_1P_2, \ldots, P_n$ is defined as a decreasing function of $t_{i+1} - t_i$ and $1/\sigma_t$ where t_{i+1} is the current time and t_i is the previous time and $1/\omega_t$ such that if $t_{i+1} - t_i \le \sigma_t$, then the weight should decrease slowly, while if $t_{i+1} - t_i > \sigma_t$ then the weight should decrease faster. The speed of the decreasing depends on the time interval $\omega_t - \sigma_t$, i.e.: a higher value results in a slower decrease, while a small value results in a faster decrease of the time-closeness weight. The set of obtained patterns were ranked according to the time-closeness weight (Definition 3) thus the closer the itemsets, the higher the measure's value is (Railean et al. 2013). Based on the Definition 3, it is possible to rank frequently changed item based on the time-closeness weight measure.

3.1 Association Rules

Association rule is the use of if/then statements to find a relationship between seemingly unrelated data or information repository (Shrivastava and Panda 2014). Association rules are split into two stages:

1. First, find all frequent itemset for predetermined minimum support (Shorman and Jbara 2017). A minimum support is set so that all itemsets with support value above the minimum support threshold are considered as frequent (Qodmanan et al. 2011). The idea of setting a support threshold is to assume that all items in the frequent data set have similar frequencies (Dunham et al. 2001). However, in reality, some items may be more frequent than other items.
2. Second, generate high confidence rules from each frequent itemset (Shorman and Jbara 2017). Rules that satisfy the minimum confidence threshold are extracted to represent the frequent itemsets. Thus, frequent itemsets are the itemsets with frequency greater than a specified threshold (Sudhir et al. 2012). Shorman and Jbara (2017) indicated that the overall performance of mining association rules is determined by the first stage. The reason is that the minimum support measure constitutes the initial stage for rules to be mined because it defines the initial characteristics of all itemsets. Therefore, any item that meets this minimum support is then considered for the next stage and thus, it determines the performance of rules.

The support of the rule is the probability Pr of an item A and B occurring together in an instance, that is Pr(A and B). The confidence of a rule measures the strength of rule implication in terms of the percentage. Hence, rules that have a confidence greater than a user-specified confidence are said to have minimum confidence. Railean et al. (2013) indicated that rules can be grouped into simple rules and complex rules.

Simple rules are of the form $V_i \rightarrow V_n$ and complex rules are of the form $V_1V_2 \ldots V_{n-1} \rightarrow V_n$ (Railean et al. 2013). In this instance, all rules of the simple form $V_i \rightarrow V_{n,}$ were combined between all V_i itemsets. For example, having the rules $A \rightarrow Y$, $B \rightarrow Y$, and $C \rightarrow Y$, a complex rules is derived as $AB \rightarrow Y$, $AC \rightarrow Y$, $BC \rightarrow Y$, $ABC \rightarrow Y$.

Srikant and Agrawal (1996) indicated that an association rule problems (that is finding the relationship between items) do not only relies on number of attributes but on numeric value of each attribute. When the number of attributes and numeric value for each attribute is combined, it could increase the complexity of search for association rules. Srikant and Agrawal (1996) approach to reduce the complexity search for association rules with a large domain is to group similar attributes together and to consider each collectively. For instance, if attributes are linearly ordered, then numeric values may be grouped into ranges. The disadvantage of this approach is that it does not work well when applied to interval data as some intervals which do not contain any numeric values are included. The Equi-depth Interval method solves this disadvantage using the depth (support) of each partition which is determined by partial completeness level. Srikant and Agrawal (1996) explained the partial completeness as set of rules obtained by considering all ranges over both raw values and partitions of the quantitative attributes. Srikant and Agrawal (1996) proposed the partial completeness measure to decide on whether an attribute of frequent item is to be partitioned or not and also the number of partition that should be required. Miller and Yang (1997), proposed the distance-based interval technique that is based on the idea that intervals that include close data values are more meaningful than intervals involving distant values.

In frequent item mining, frequent items are mostly associated with another but many of these items are meaningless which subsequently generate many useless rules. A strategy to avoid meaningless items was identified by Han and Fu (1995) by splitting the data into groups according to the support threshold of the items and then discovers association rules in each group with a different support threshold. Thereafter, a mining algorithm is applied to find the association rules in the numerical interval data. This approach helps to eliminate information that does not appear enough times in the data set (Dunham et al. 2001). Dunham et al. (2001) indicated that when items are grouped into a few blocks of frequent items, a single support threshold for the entire data set is inadequate to find important association rules as it cannot find inherent differences in frequency of items in the data set. Grouping frequent items may require partitioning of the numeric value attributes into intervals. The problem is that, when the number of values/intervals for an attribute is very large, the support value of particular values/intervals may be low. Again, if the number of values/intervals for an attribute is small, there is a possibility of losing information. That means, some rules may not have a threshold for confidence value. In order to overcome aforementioned problems, all possible ranges over values/intervals may be combined when processing each particular value/interval (Dunham et al. 2001). Dunham et al. (2001) proposed three steps in solving the problem of finding frequent itemsets with quantitative attributes. At the first step, decide whether each attribute is to be partitioned or not. If an attribute is to be partitioned, determine the number of partitions. During the second step, map the values of the attribute to a set of consecutive integers. During the third step, find the support of each value of all attributes. In order to avoid minimum support problem, adjacent values are combined as long as their support is less than user-specified maximum support (Dunham et al. 2001). However, during partitions, some information might be lost and this information loss is measured in terms of how close rules are. For instance, if R is a set of rules obtained over raw values and $R1$ is set of rules over the partition of quantitative attributes then the closeness is the difference between R and $R1$. A close rule is found if the minimum confidence threshold for $R1$ is less than minimum confidence for R by a specified value.

Although, the traditional methods such as support and confidence formulation are useful in extracting association rules from both attributes and numeric value dimension, it does not consider the time closeness between antecedent and consequent of rules. Time closeness may be significant because it helps in knowing how time plays a role in determining the number of items in a rule.

3.2 Apriori Algorithm

Apriori algorithm is a well-known algorithm used in data mining for discovering association rules. The basic concept about the Apriori is that if an itemset is frequent, then all of its subsets must also be frequent (Agrawal and Srikant 1994a, b) so as to find all frequent items. This basic concept/property enables the Apriori algorithm to efficiently generate a set of candidate large itemsets whose lengths are $(k + 1)$ from the large k-itemsets (for $k \geq 1$) and eliminates candidates which do not contain large subsets. Thus, if the candidate itemsets satisfies minimum support, then it is frequent itemsets while, for the remaining candidates, only those with support over minimum support threshold is taken to be large $(k + 1)$-itemsets.

In Apriori algorithm, search strategy should help in pruning itemsets (Geng and Hamilton 2006) and the search strategy for is based both on the breadth-first search and on a tree structure to count candidate itemsets (Shorman and Jbara 2017). During counting of itemset, only the frequent itemsets, found in the previous pass, are used because it fulfils the property indicated earlier. Although the Apriori can generate frequent itemsets, an aspect which is yet to be considered is how to discover association rules on frequently changed itemsets with a time dimension. The proposed algorithm comprises of two parts, preprocessing and mining. The pre-processing part calculates the fitness values of each kestrel. The support of the fitness function is expressed using the Support-Closeness Preference-based measure combined with an additional weighting function to fulfil the anti-monotone property (Railean et al. 2013), with a subsequent post-processing phase with Modified Closeness Preferences with support and confidence values (MCPsc). The mining part of the algorithm, which constitutes the major contribution of the paper, uses the KSA algorithm to mine association rules.

4 Conclusion

In this chapter, we discussed the data mining, the meta-heuristic search methods to data mining of association rules. The advantage of meta-heuristic search methods is the ability to self-tune parameters in order to fine-tune the search for rules to extract. Since time is significant in search for rules, the chapter also considered the use of closeness preferences model which ensures rules are extracted within a time interval defined by a user. As the needs of users may vary, the closeness preference helps to cater for varying user time so as to extract rules.

Key Terminology and Definitions

Data mining is the process of finding hidden and complex relationships present in data with the objective to extract comprehensible, useful and non-trivial knowledge from large data sets.

Association rule uses if/then statements to find a relationship between seemingly unrelated data. The support and confidence criteria help to identify the important relationships between items in data set.

Closeness Preference refers to time interval at which a user selects rules with closer antecedent and consequent.

References

Agbehadji, I. E. (2011). Solution to the travel salesman problem, using omicron genetic algorithm. *Case study: Tour of national health insurance schemes in the Brong Ahafo region of Ghana* (Online Master's Thesis).

Agbehadji, I. E., Millham, R., & Fong, S. (2016). Wolf search algorithm for numeric association rule mining. In: *2016 IEEE International Conference on Cloud Computing and Big Data Analysis (ICCCBDA 2016)*, Chengdu, China.

Agrawal, R., & Srikant, R. (1994a). Fast algorithms for mining association rules in large databases. In: *Proceedings 20th International Conference on Very Large Data Bases* (pp. 478–499).

Agrawal, R., & Srikant, R. (1994b). Fast algorithms for mining association rules. In: *Proceedings of the 20th International Conference on Very Large Databases*, Santiago, Chile (pp. 487–499).

Al-Ani, A. (2007). Ant colony optimization for feature subset selection. *World Academy of Science, Engineering and Technology International Journal of Computer, Electrical, Automation, Control and Information Engineering, 1*(4).

Cao, L., Luo, D., & Zhang, C. (2007). Knowledge actionability: Satisfying technical and business interestingness. *International Journal Business Intelligence and Data Mining, 2*(4).

Cheng, S., Shi, Y., Qin, Q., & Bai, R. (2013). *Swarm intelligence in big data analytics*. Berlin: Springer.

Dorigo, M., Birattari, M., & Stützle, T. (2006). Ant colony optimization: Artificial ants as a computational intelligence technique.

Dorigo, M., Maniezzo, V., & Colorni, A. (1996). Ant system: Optimization by a colony of cooperating agents. *IEEE Transactions on Systems, Man, and Cybernetics, Part B (Cybernetics), 26*(1), 29–41.

Dunham, M. H., Xiao, Y., Gruenwald, L., & Hossain, Z. (2001). A survey of association rules.
Fister, I., Jr., Fong, S., Bresta, J., & Fister, I. (2014). Towards the self-adaptation of the bat algorithm. In: *Proceedings of the IASTED International Conference, February 17–19, 2014 Innsbruck, Austria Artificial Intelligence and Applications.*
Fung, B. C. M., Wang, K., & Liu, J. (2012). Direct discovery of high utility itemsets without candidate generation. In: *2012 IEEE 12th International Conference on Data Mining.*
Garcia, S., Luengo, J., & Herrera, F. (2015). Data preprocessing in data mining. In *Intelligent systems reference library* (Vol. 72). Springer International Publishing Switzerland. https://doi.org/10.1007/978-3-319-10247-4_3.
Geng, L., & Hamilton, H. J. (2006). Interestingness measures for data mining: A survey. *ACM Computing Surveys, 38*(3), Article 9. (Publication date: September 2006).
Han, J., Cheng, H., Xin, D., & Yan, X. (2007). Frequent pattern mining: current status and future directions. *Data Mining Knowledge Discovery, 15,* 55–86.
Han, J., & Fu, Y. (1995). Discovery of multiple-level association rules from large databases. In: *Proceedings of the 21st International Conference on Very Large Databases*, Zurich, Swizerland (pp. 420–431).
Heraguemi, K. E., Kamel, N., & Drias, H. (2015). Association rule mining based on bat algorithm. *Journal of Computational and Theoretical Nanoscience, 12,* 1195–1200.
Jagtap, S., Kodge, B. G., Shinde, G. N., & Devshette P. M. (2012). Role of association rule mining in numerical data analysis. *World Academy of Science, Engineering and Technology International Journal of Computer, Electrical, Automation, Control and Information Engineering, 6*(1).
Kennedy, J., & Eberhart, R. C. (1995). Particle swarm optimization. In: *Proceedings of IEEE International Conference on Neural Networks*, Piscataway, NJ, pp 1942–1948.
Krause, J., Cordeiro, J., Parpinelli, R. S., & Lopes, H. S. (2013). A survey of swarm algorithms applied to discrete optimization problems.
Kuo, R. J., Chao, C. M., & Chiu, Y. T. (2011). Application of PSO to association rule. *Applied Soft Computing.*
Kuo, R. J., & Shih, C. W. (2007). Association rule mining through the ant colony system for National Health Insurance Research Database in Taiwan. *Computers & Mathematics with Applications, 54*(11–12), 1303–1318.
Mannila, H., Toivonen, H., & Verkamo, A. I. (1997). Discovery of frequent episodes in event sequences. *Data Mining and Knowledge Discovery, 1*(3), 259–289.
Miller, R.J., & Yang, Y. (1997). Association rules over interval data. In: *SIGMOD 1997, Proceedings ACM SIGMOD International Conference on Management of Data*, 13–15 May 1997, Tucson, Arizona, USA (pp. 452–461). ACM Press.
Olmo, J. L., Luna, J. M., Romero, J. R., & Ventura, S. (2011). Association rule mining using a multi-objective grammar-based ant programming algorithm. In: *2011 11th International Conference on Intelligent Systems Design and Applications* (pp. 971–977). IEEE.
Palshikar, G. K., Kale, M. S., & Apte, M. M. (2007). *Association rules mining using heavy itemsets.*
Piatetsky-Shapiro, G. (1991). Discovery, analysis and presentation of strong rules. In G. Piatetsky-Shapiro & W. J. Frawley (Eds.), *Knowledge Discovery in Databases* (p. 229). AAAI.
Qodmanan, H. R., Nasiri, M., & Minaei-Bidgoli, B. (2011). Multi objective association rule mining with genetic algorithm without specifying minimum support and minimum confidence. www.elsevier.com/locate/eswa.
Railean, I., Lenca, P., Moga, S., & Borda, M. (2013). Closeness-preference—A-new-interestingness-measure-for-sequential-rules-mining. *Knowledge-Based-Systems.*
Rajaraman, A., & Ullman, J. (2011). *Mining of massive datasets.* New York: Cambridge University Press.
Sarath, K. N. V. D., & Ravi, V. (2013). Association rule mining using binary particle swarm optimization. *Engineering Applications of Artificial Intelligence.* www.elsevier.com/locate/engappai.

Shorman, H. M. A., & Jbara, Y. H. (2017, July). An improved association rule mining algorithm based on Apriori and Ant Colony approaches. *IOSR Journal of Engineering (IOSRJEN)*, 7(7), 18–23. ISSN (e): 2250-3021, ISSN (p): 2278-8719.

Shrivastava, A. K., & Panda, R. N. (2014). Implementation of Apriori algorithm using WEKA. *KIET International Journal of Intelligent Computing and Informatics, 1*(1).

Silberschatz, A., & Tuzhilin, A. (1995). On subjective measures of interestingness in knowledge discovery. *Knowledge Discovery and Data Mining*, 275–281.

Song, M., & Rajasekaran, S. (2006). A transaction mapping algorithm for frequent itemsets mining. *IEEE Transactions on Knowledge and Data Engineering, 18*(4), 472–481.

Srikant, R., & Agrawal, R. (1996). Mining quantitative association rules in large relational tables. In: *Proceedings of the 1996 ACM SIGMOD International Conference on Management of Data*, Montreal, Quebec, Canada (pp. 1–12), 4–6 June 1996.

Stützle, T., & Dorigo, M. (2002). *Ant colony optimization*. Cambridge, MA: MIT Press. https://pdfs.semanticscholar.org/7c72/393febe25ef5ce2f5614a75a69e1ed0d9857.pdf.

Sudhir, J., Kodge, B. G., Shinde, G. N., & Devshette P. M. (2012). Role of association rule mining in numerical data analysis. *World Academy of Science, Engineering and Technology International Journal of Computer, Electrical, Automation, Control and Information Engineering, 6*(1).

Sumathi, S., & Sivanandam, S. N. (2006). Introduction to data mining principles. *Studies in Computational Intelligence (SCI), 29*, 1–20. www.springer.com/cda/content/…/cda…/9783540343509-c1.pdf.

Tang, R., Fong, S., Yang, X-S, & Deb, S. (2012). *Wolf search algorithm with ephemeral memory*. IEEE.

Tseng, V. S, Liang, T. and Chu, C (2006), Efficient Mining of Temporal High Utility Itemsets from Data streams. UBDM'06, August 20, 2006, Philadelphia, Pennsylvania, USA.

Wang, K., Zhou, S., & Han, J. (2002). Profit mining: from patterns to actions. In: *EBDT 2002*, Prague, Czech (pp. 70–87).

Wei, Y., Huang, J., Zhang, Z., & Kong, J. (2015). SIBA: A fast frequent item sets mining algorithm based on sampling and improved bat algorithm.

Wu, C., Buyya, R., Ramamohanarao, K. (2016). Big Data Analytics = Machine Learning + Cloud Computing. arXiv preprint arXiv:1601.03115.

Yang, X.-S. (2008). *Nature-inspired metaheuristic algorithms*. Luniver Press.

Yang, X. S. (2009). Firefly algorithm, Levy flights and global optimization. In: *XXVI Research and Development in Intelligent Systems*. Springer, London, UK, pp 209–218.

Yang, X. (2010). A new metaheuristic bat-inspired algorithm. In: *Nature Inspired Cooperative Strategies for Optimization (NICSO 2010)*. Springer, pp. 65–74.

Richard Millham is currently an Associate Professor at Durban University of Technology in Durban, South Africa. After thirteen years of industrial experience, he switched to academe and has worked at universities in Ghana, South Sudan, Scotland, and Bahamas. His research interests include software evolution, aspects of cloud computing with m-interaction, big data, data streaming, fog and edge analytics, and aspects of the Internet of things. He is a Chartered Engineer (UK), a Chartered Engineer Assessor and Senior Member of IEEE.

Israel Edem Agbehadji graduated from the Catholic University college of Ghana with B.Sc. Computer Science in 2007, M.Sc. Industrial Mathematics from the Kwame Nkrumah University of Science and Technology in 2011 and Ph.D. Information Technology from Durban University of Technology (DUT), South Africa, in 2019. He is a member of ICT Society of DUT Research group in the Faculty of Accounting and Informatics; and *IEEE* member. He lectured undergraduate courses in both DUT, South Africa, and a private university, Ghana. Also, he supervised several undergraduate research projects. Prior to his academic career, he took up various managerial positions as the management information systems manager for National Health Insurance Scheme and the postgraduate degree programme manager in a private university in Ghana.

Currently, he works as a Postdoctoral Research Fellow, DUT-South Africa, on joint collaboration research project between South Africa and South Korea. His research interests include big data analytics, Internet of things (IoT), fog computing and optimization algorithms.

Hongji Yang graduated with a Ph.D. in Software Engineering from Durham University, England with his M.Sc. and B.Sc. in Computer Science completed at Jilin University in China. With over 400 publications, he is full professor at the University of Leicester in England. Prof. Yang has been an IEEE Computer Society Golden Core Member since 2010, an EPSRC peer review college member since 2003, and Editor in Chief of the International Journal of Creative Computing.

Chapter 6
Lightweight Classifier-Based Outlier Detection Algorithms from Multivariate Data Stream

Simon Fong, Tengyue Li, Dong Han, and Sabah Mohammed

1 Introduction

A large number of outlier-detection-based applications exist in data mining, such as credit card fraud detection in financial field, clinical trials observation in medical field, voting irregularity analysis in sociology field, data cleansing, intrusion detection system in computer networking domain, severe weather prediction in meteorology, geographic information system in geology, and athlete performance analysis in sports field. The list goes on for many other possible data mining tasks.

In the era of big data, we are facing two problems in the perspective of big data analytics. It is known that traditionally outlier detection algorithms work with the full set of data. Outliers are computed in relation between some extraordinary data and the rest of the data which is in reference to the whole set of data. Nowadays, with the advances of data collection technologies, data are often generated in data streams. The data are produced in sequences of data stream that demand for new data mining algorithms that are able to incrementally learn or process the data stream without the need of loading in the full data when new data arrives. Outlier detection which is a member of data mining family has no exception. Upon working with big data, it is good to have an outlier detection algorithm that rides over the data stream, and by using some suitable statistical measures to find outliers on the fly.

S. Fong (✉) · T. Li · D. Han
Department of Computer Science, University of Macau, Taipa, Macau SAR
e-mail: ccfong@umac.mo

T. Li
e-mail: litengyue2018@gmail.com

S. Mohammed
Department of Computer Science, Lakehead University, Thunder Bay, Canada
e-mail: sabah.mohammed@lakeheadu.ca

S. J. Fong and R. C. Millham (eds.), *Bio-inspired Algorithms for Data Streaming and Visualization, Big Data Management, and Fog Computing*, Springer Tracts in Nature-Inspired Computing,
https://doi.org/10.1007/978-981-15-6695-0_6

Given this context of incremental detection of outliers over a data stream that potentially can amount to infinity, several computational challenges exist: (1) a dataset that is continuously generated and comprised or merged from data of various sources would contain many attributes; (2) the outlier detection algorithm must have certain satisfactory accuracy and reasonably fast time. The detection rate must be equal to or higher than the data generation speed; and (3) the incremental outlier detection algorithm should be adaptive to the incoming data stream at any stage of time. Ideally, it should learn about the patterns and the characteristics of any part of the data stream, as they are prone to change from time to time. Hence, we look into the concept of classifier-based outlier detection algorithm.

A dual-step solution is formulated to meet the challenges as aforementioned. On one hand, data stream is processed by loading a sliding window of data in real time, one at a time, instead of loading the whole data before we apply the outlier detection algorithms. On the other hand, some suitable arithmetic methods are needed to calculate the value of Mahalanobis or local outlier factor as a measure of how far the data point is from the central. Furthermore, an effective interquartile range (IQR) method is used in conjunction to formulate our proposed classifier-based outlier detection (COD). The outlier algorithm has to satisfy the constraints of time or speed and accuracy because data is streaming in fast.

This study reported in this chapter is about investigating the performance for outlier detection from multivariate data stream, and how the performance under the condition of incremental processing could be improved by using COD. IQR is first treated as a preprocessing step, which filters the data in unsupervised manners. Then, lightweight method can be applied to most of the outlier detection situations, regardless the dataset is continuously generated data stream or disposable loading of multivariate data. Lightweight method is also an user-defined algorithm, because it allows user to specify any parameters (e.g., the boundary value and degree of confidence) to fit for the target data stream. Finally, a classification algorithm will be applied.

In the rest of the chapter, we would organize the contents as follows: A literature survey over is presented, which introduces the previous works of the outlier detection. In the following section, the advantages of the proposed COD are presented. The experiments which are conducted as a comparative study follow. For discussing the results, charts and diagrams of some "important technical indicators (e.g., time spend, correctly classified instances, Kappa statistic, mean absolute error, root-mean-squared error, TP rate) are shown so as to reinforce the efficacy of our novel approach." At the end, a conclusion which remarks and lists out the future works is given.

In outlier detection, we can use many detection indices to find out those abnormal values, like LOF value or Mahalanobis. What we should not ignore is that these indices are all calculated after mathematical operation. That means these values are generated from some complicated formula. Sometime, we should aware that another outlier detection direction, which is called a statistical operation. These two operations do not have conflicts with each other (Fig. 1).

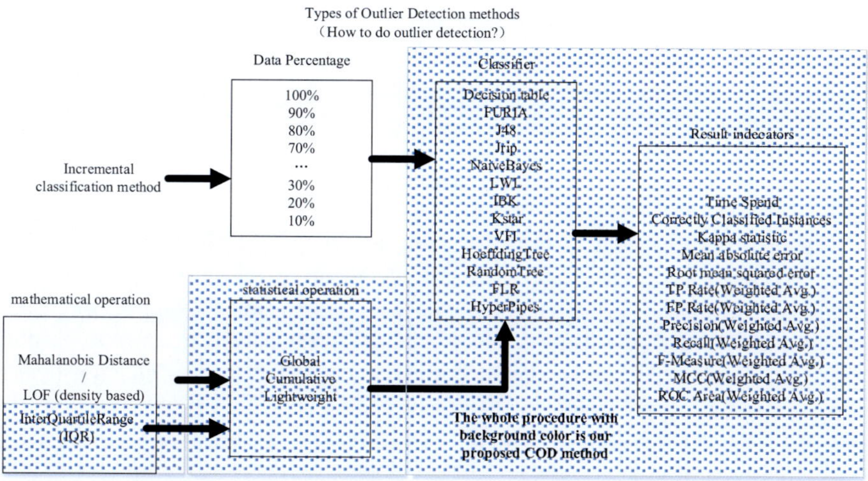

Fig. 1 Overall flow diagram

1.1 *Incremental Classification Method*

The incremental classification method is set as a control group in our experiment. Firstly, we cut each datasets by different proportions of original dataset into 10 copies. In particular, we resample the data to form 100% of the raw data to 90, 80, 70, 60, 50, 40, 30, 20, 10% of the raw data by the filter "ReservoirSample." After 9 times of resample, we have 9 new datasets, or 10 dataset which includes the original dataset.

Then, for each classification algorithm, we used all these 10 copies to get the evaluation results and recorded them in a table.

Finally, we use the coordinate systems to reflect the relationship among sampling rate, accuracy rate, and time spending. From the comparison of two x axes and one y-axis, we can easily observe the disadvantages and advantages among classification algorithms.

In a word, this method cuts the original data into many parts, ranging from small to big. This increment rule for instances is what we call "incremental." In practical, most of the classification algorithms could be applied in the incremental way if we load the data instances not all at once.

1.2 *Mahalanobis Distance with Statistical Methods*

The method is set as a control group as well. In terms of the earliest statistical-based outlier detection method, this method can only be applied to single dimensional datasets, notably, datasets of univariate outliers. We could compute the "Z-score" for each number and compared it with a predefined threshold. Thus, a positive standard score indicates a datum above the mean, while a negative standard score indicates a datum below the mean.

However, "in practice, we usually encounter more complex situations with multi-dimensional records." One method that can be used for recognizing multivariate outliers is termed as Mahalanobis distance. It measures the distance of particular scores from the centroids (denoted as P) of the remaining samples (denoted as D). This "distance is zero if P is at the mean of D, and grows as P moves away from the mean: Along each principal component axis, it measures the number of standard deviations from P to the mean of D." If each of "these axes is rescaled to have unit variance, then Mahalanobis distance corresponds to standard Euclidean distance in the transformed space." Mahalanobis distance is thus "unit less and scale-invariant and takes into account the correlations of the dataset."

From the aspect of the definition, the Mahalanobis distance of a vector x which is multidimensional is $(x_1, x_2, x_3, \ldots, x_N)^{\mathrm{T}}$ from a collections of data values with mean $\mu = (\mu_1, \mu_2, \mu_3, \ldots \mu_N)^{\mathrm{T}}$ and covariance matrix S is defined as:

$$D_M(x) = \sqrt{(x-u)^{\mathrm{T}} S^{-1} (x-u)} \quad (1)$$

We define each x is the Mahalanobis distance score from the reference sample, while the u is the mean of the specific reference sample and covariance S is the covariance of the data in reference sample. According to the algorithm of Mahalanobis distance, the quantity of instances in reference sample must be greater than the quantity of the dimension.

For multivariate data that are normally distributed, they can be approximately chi-square distributed with p degrees of freedom (x_p^2). Multivariate outliers can now easily be defined as observations having a large (squared) Mahalanobis distance. "After calculating the Mahalanobis distance for a multivariate instance from the specific data group, we will get a squared Mahalanobis distance score." If this score exceeds a "critical value," this instance will be considered an outlier. When $p < 0.05$, we generally refer to this as a significant difference.

For example, the critical value for a bivariate relationship is 13.82. Any "Mahalanobis distances score above that critical value is a bivariate outlier." In our experiment, we will use the Mahalanobis distance as an indicator to find the outliers.

For the global analysis, we calculate the "Mahalanobis Distance for every instance in the whole dataset." This method is similar to the traditional one, which loaded the whole data at once. The following "diagrams indicate the Global Analysis mechanism with Mahalanobis Distance of the ith and its next instance." The "operation of global analysis using MD is visualized" in Fig. 2a.

For the "cumulative analysis, in our experiment, initially we calculate the Mahalanobis distance of the first 50 records, respectively." After that, "for the ith record,

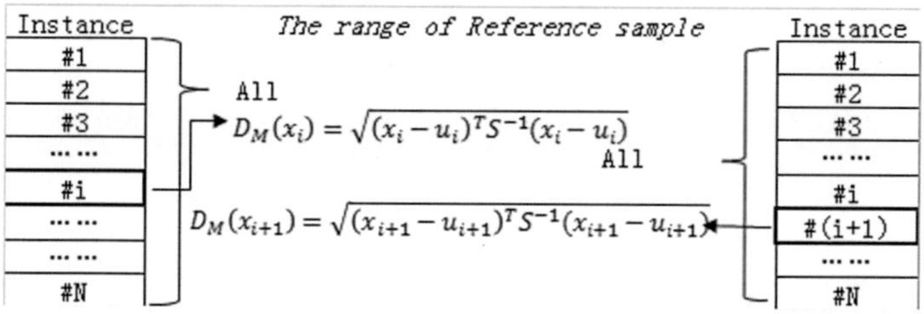

(a): MD Global Analysis

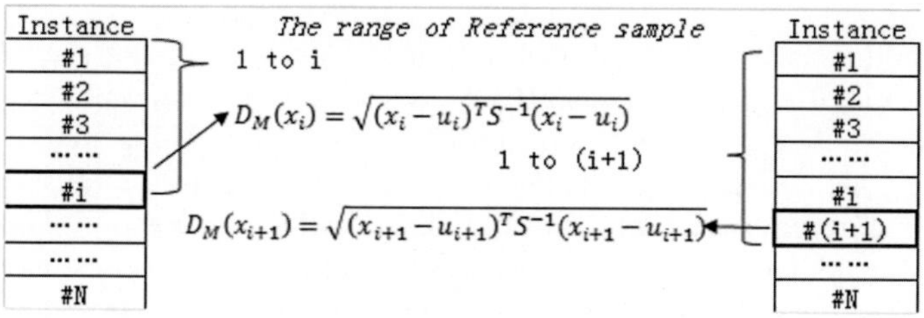

(b): MD Cumulative Analysis

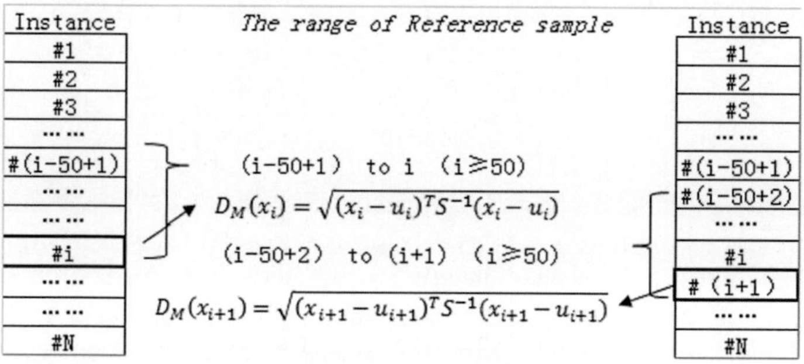

(c): MD Lightweight Analysis

Fig. 2 Outlier detection using MD with different operation modes

we treat the top i instances as the reference sample." The following diagrams indicate the "Cumulative Analysis mechanism for calculating the Mahalanobis Distance" of the ith and its next instance. The operation of "cumulative analysis using MD is visualized" in Fig. 2b.

As to the "lightweight analysis with sliding window, we propose a novel notion which called sliding window." The "sliding window has a fixed size of a certain number of instances; 100 in our experiments, and it moves forward to next instance when we analyze a new record." For example, if we choose the window size of 50, each record within it will compute the "Mahalanobis Distance from the reference sample (namely, the selected 1–50 instances)." Then, the window will slide forward by a step of one record. So, the window is formed with the instances of 2–51. That is to say, we calculate the 51st instance's "Mahalanobis Distance from the reference sample formed by records from 2 to 51. The following diagrams indicate the method about lightweight analysis with window size of 50 for calculating the Mahalanobis distance of the *i*th and its next instance." The operation of "lightweight analysis using MD is visualized" in Fig. 2c.

1.3 Local Outlier Factor with Statistical Methods

The method is set as a control group as same. "Outlier ranking is a well-studied research topic. Breunig et al. (2000) have developed the local outlier factor (LOF) system that is usually considered a state-of-the-art outlier ranking method. The main idea of this method is to try to obtain a divergence score for each instance by estimating its degree of isolation compared to its local neighborhood." The method is based on the notion of the "local density of the observations." Cases in regions with very low density are considered outliers. "The estimates of the density are obtained using the distances between cases." The authors defined a few "concepts that drive the algorithm used to calculate the divergence score of each point. These are the (1) concept of core distance of a point *p*, which is defined as its distance to its *k*th nearest neighbor, (2) concept of reachability distance between the case $p1$ and p_2, which is given by the maximum of the core distance of $p1$ and the distance between both cases, and (3) local reachability distance of a point, which is inversely proportional to the average reachability distance of its *k* neighbors." The "LOF of a case is calculated as a function of its local reachability distance." In addition, there are "2 parameters that denote the density. One parameter is *MinPts* that controls the minimum number of objects and the other parameter specifying a volume." These "2 parameters determine a density threshold for the clustering algorithms to operate." That is, "objects or regions are connected if their neighborhood densities exceed the predefined density threshold."

In [7], the author "summarized the definition of local outlier factor as follows":

"Let *D* be a database."

"Let *p, q, o* be some objects in *D*."

"Let *k* be a positive integer."

We "use $d(p, q)$ to denote the Euclidean distance between objects *p* and *q*. To simplify our notation, $d(p, D) = min\{d(p, q) \mid q \in D\}$. Based on the above assumptions, we define the following five definitions."

(1) Definition 1: *k*-distance of *p*

"The k-distance of p, denoted as k-distance(p) is defined as the distance d(p; o) between p and o such that:

(i) *for at least k objects* $o' \in D \setminus \{p\}$ *it holds that* $d(p,o') \leq d(p,o)$, *and*
(ii) *for at most k-1 objects* $o' \in D \setminus \{p\}$ *it holds that* $d(p,o') < d(p,o)$.*"*

Intuitively, k-distance(p) "provides a measure on the sparsity or density around the object p. When the k-distance of p is small, it means that the area around p is dense and vice versa."

(2) Definition 2: "k-distance neighborhood of p

The k-distance neighborhood of p contains every object whose distance from p is not greater than the k-distance, is denoted as"

$N_k(p) = \{q \in D\backslash\{p\} \mid d(p, q) \leq k\text{-distance}(p)\}$.

Note "that since there may be more than k objects within k-distance(p), the number of objects in $N_k(p)$ may be more than k." Later on, the "definition of LOF is introduced, and its value is strongly influenced by the k-distance of the objects in its k-distance neighborhood."

(3) Definition 3: "reachability distance of p w.r.t object o

The reachability distance of object p with respect to object o is defined as

$\text{reach-dist}_k(p, o) = \max\{k\text{-distance}(o), d(p, o)\}$."

(4) Definition 4: "local reachability density of p

The local reachability density of an object p is the inverse of the average reachability distance from the k-nearest-neighbors of p."

$$lrd_k(p) = \frac{|N_{k(p)}(p)|}{\sum o \in N_k(p) reach - dist_k(p, o)} \tag{2}$$

"Essentially, the local reachability density of an object p is an estimation of the density at point p by analyzing the k-distance of the objects in $N_k(p)$." The local "reachability density of p is just the reciprocal of the average distance between p and the objects in its k-neighborhood." Based on local reachability density, the local outlier factor can be defined as follows.

(5) Definition 5: local outlier factor of p

$$\text{LOF}_k(p) = \frac{\sum o \in N_k(p) \frac{\text{lrd}(o)}{\text{lrd}(p)}}{|N_k(p)|} \tag{3}$$

"LOF is the average of the ratios of the local reachability density of p and those of p's k-nearest-neighbors. Intuitively, p's local outlier factor will be very high and its local reachability density is much lower than those of its neighbors." "A value of approximately 1 indicates that the object is comparable to its neighbors (and thus not an outlier)." "A value below 1 indicates a denser region (which would be an inlier), while values significantly larger than 1 indicate outliers."

In our experiment, we set "the inspection effort to 0.1 which means that we regard the top 10% records as outliers according to the outlier score in decreasing sequence."

For the global analysis, we calculate the "LOF score for each instance from the whole dataset." The following diagrams indicate the "Global Analysis mechanism for calculating the LOF score of the *i*th and its next instance." The operation of global analysis using LOF is visualized in Fig. 3a.

"As to the cumulative analysis, in our experiment, at first we calculate the LOF scores of the first 50 records, respectively, and labeled top 10% of the highest score ones as outliers." After that, for the *i*th record, "calculate the LOF score for all these i records, and then examine this *i*th one to see whether it is among the top 10% highest score of the present dataset." "If yes, then this instance is regarded as an outlier." "Otherwise, it is normal." The following diagrams indicate the "Cumulative

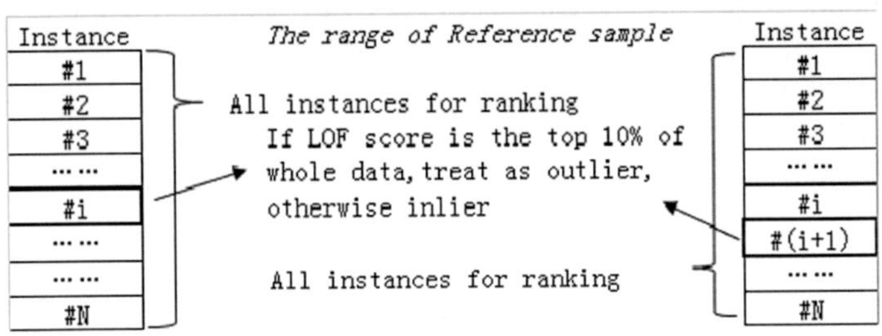

(a) LOF Global Analysis

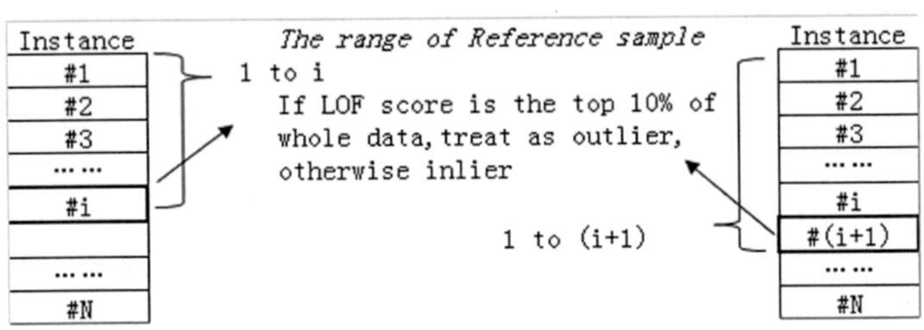

(b) LOF Cumulative Analysis

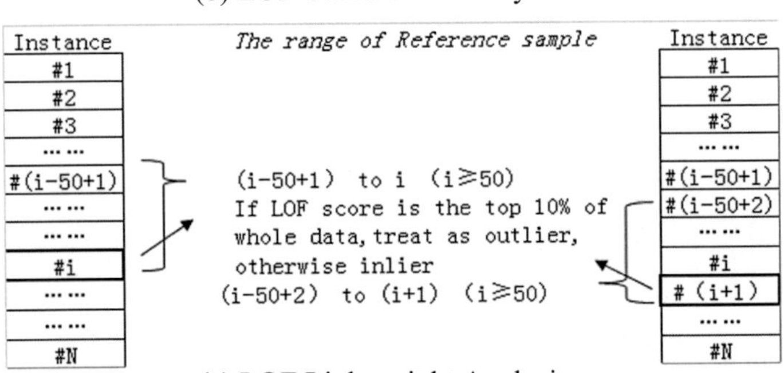

(c) LOF Lightweight Analysis

Fig. 3 Outlier detection using LOF with different operation modes

Analysis mechanism for calculating the LOF score of the ith and its next instance." The operation of cumulative analysis using LOF is visualized in Fig. 3b.

"The mechanism of lightweight analysis with LOF method to detect outliers is similar to the mechanism of Mahalanobis distance method" which is mentioned above. The following diagram in Fig. 3c indicates the method about "Lightweight Analysis using LOF with window size of 50 to estimate an outlier of the ith and its next instance."

1.4 Classifier-Based Outlier Detection (COD) Methods

"The method is set as a test group. The core step for COD is to calculate the IQR value for each instance at the very beginning." The "interquartile range (IQR), or called middle fifty as well is the concept in descriptive statistics." It is a measure of "statistical dispersion, being equal to the difference between the upper and lower quartiles." In practical to find outliers in data, we define outliers are those observations that fall below $Q1 - 1.5(\text{IQR})$ or above $Q3 + 1.5(\text{IQR})$ (Fig. 4).

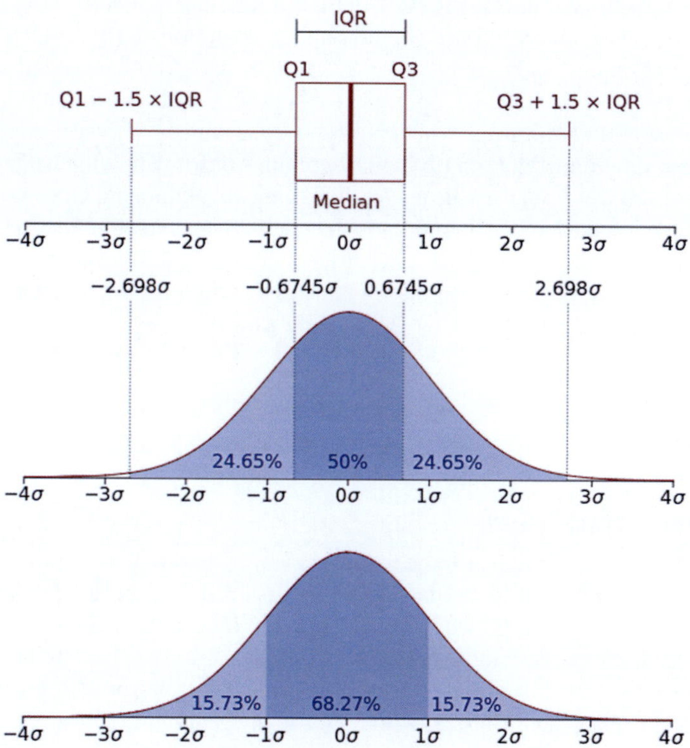

Fig. 4 Boxplot and a probability density function of a Normal $N(0, \sigma 2)$ population

We should say, no matter what statistical method we used (global, cumulative, lightweight), Mahalanobis distance, local outlier factor, or interquartile range are just a prerequired value. Only to get these results, we could use the statistical method to judge whether it is an abnormal value. In practice, although in many of the results accuracy is high, they spent a lot of time in preprocessing.

Then, in order to achieve consistency of comparison, we also have a total of 3 workflows for this COD method. These workflows are (1) global analysis using classifier-based outlier detection method, (2) cumulative analysis using classifier-based outlier detection method, and (3) lightweight analysis using classifier-based outlier detection method. The global workflow is also called "traditional," while another two are collectively referred to as incremental approach.

For the (1) global analysis using classifier-based outlier detection, we apply the percentage spilt option in the "classify" tab. The parameter we use in this dataset is 64, which means 64% of the instances are used as test data. The target class should be "outlier" in the last column. Before applying each classification algorithm, we should set the test option as above mentioned.

For the (2) cumulative analysis using classifier-based outlier detection method, we apply the "cross-validation" option in the "classify" tab. The file we use in this dataset 1000 instances, which is generated by function "reservoir sampling" from the original data. The target class should also be "outlier" in the last column. Before applying each classification algorithm, we should set the test option as above mentioned.

For the (3) lightweight analysis using classifier-based outlier detection method, we apply the "supplied test set" option in the "classify" tab. The fold number we use in this dataset 1000 instances, which is first 1000 instances of the original data. The target class should be "outlier" in the last column. Before applying each classification algorithm, we should set the test option as above mentioned. Our experiment of outlier detection pays much attention on the comparison and combination of mathematic approach and statistical approach, especially IQR and lightweight with sliding window. We should choose the most suitable association for the target dataset type. Difficult issues and challenges lay on this field as well.

2 Proposed Methodology

2.1 Data Description

The "UCI Machine Learning Repository is a collection of databases, domain theories, and data generators that are used by the machine learning community for the empirical analysis of machine learning algorithms."

Hence, we select two datasets, which is data 1 and data 3, from this impact archive to conduct the experiments. Meanwhile, we use the generator to generate the second dataset (data 2) as well.

In the dataset, "Statlog (Shuttle)" contains "9 attributes and 58,000 instances; we choose 43,500 continuous instances of the whole." The examples in the "original dataset were in time order, and this time order could presumably be relevant in classification." However, this was not deemed relevant for Statlog purposes, so the order of the examples in the original dataset was randomized, and a portion of the original dataset removed for validation purposes.

We also use another two datasets, which is data 2 and data 3, to conduct the experiments and get the results for part 3.2, 3.3 and 3.4 of this study.

In the second "Random" dataset contains 5 attributes and 10,000 instances. These data are automatically generated from the "generate" function in "Preprocess" tab in WEKA. The generator is RDG1 with the default settings, expect for the numAttributes and numExamples.

In the third dataset (data 3), all data are from one continuous EEG (Electroencephalogram, a test or record of brain activity produced by electroencephalography) measurement with the Emotivss EEG Neuroheadset. The duration of the measurement was 117 s. The eye state was detected via a camera during the EEG measurement and added later manually to the file after analyzing the video frames. '1' indicates the eye-closed and '0' the eye-open state. All values are in chronological order with the first measured value at the top of the data. This dataset contains 15 attributes and 14,980 instances.

2.2 Comparison Among Classification Algorithms

In our study, we applied the classification algorithms to test the final outlier detection. These algorithms are already embedded in WEKA environments, which includes decision table, FURIA (Fuzzy Rule Induction algorithm), HoeffdingTree, IBK, J48, LWL, JRip, K-Star, VFI, Naive Bayes, Random Tree. These algorithms are used to verify the improvement of the proposed method. All of these classification algorithms are set with the default values for each algorithm and using the 10 folds cross-validation as the test option. The last attribute of each dataset is the target class. Hence, we need to apply these algorithms in sequence and get different components, like Kappa statistic, from the result.

A collection of classification algorithms were tested, such as decision table, FURIA (Fuzzy Rule Induction algorithm), HoeffdingTree, IBK, J48, LWL, JRip, K-Star, VFI, Naive Bayes, and Random Tree.

3 Results and Analysis

As we proposed earlier, there could be three main categories of comparisons. These test groups are incremental classification method, Mahalanobis distance with statistical method, and local outlier factor with statistical method. All these methods belong to test group would compare to our proposed COD method.

3.1 Results for Incremental Classification Methods

This experiment is conducted by scaling down the sampling rate when training the dataset. For example, 100 means the full dataset is being used. That is highest accuracy we could have because of the full data. However, the cost of model bulling time is neglected. For simple comparison, all the tested algorithms are divided into two groups. Correctly Classified Instances, red line and axes *y*, almost on the top of each figure. While the Time Spending, blue line and axes *x*, almost on the bottom of each figure. The stable group, includes *k*-star, IBK, random tree, FURIA, JRip, Decision Table, J48 almost achieve the good accuracy to 100%, and the model building time linear decreasing with the amount of data. The "unstable group includes LWL, VFI, Hoeffding tree, Naive Bayes and does not have linear decreasing features." As the results for the chosen data, those algorithms belongs to stables group (Figs. 5, 6, 7, 8, 9, 10 and 11) outer perform those in unstable group (Figs. 12, 13, 14 and 15) some of the time. This safely reached conclusion that accuracy rate and time should be taken into account when building the mode.

In this stable group, all algorithms are nearly achieving the 100% perfect accuracy in all sampling amounts. Those algorithms belong to stable group all gains the accuracy greater than 99%. K-stars and IBK almost take no time in building the model: It takes approximately 0–0.01 s for model building, no matter the full data or the

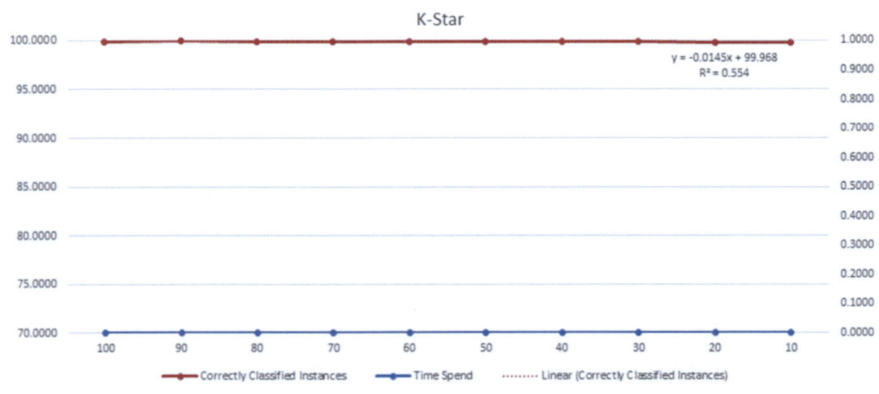

Fig. 5 Result for K-star algorithm based on incremental classification method

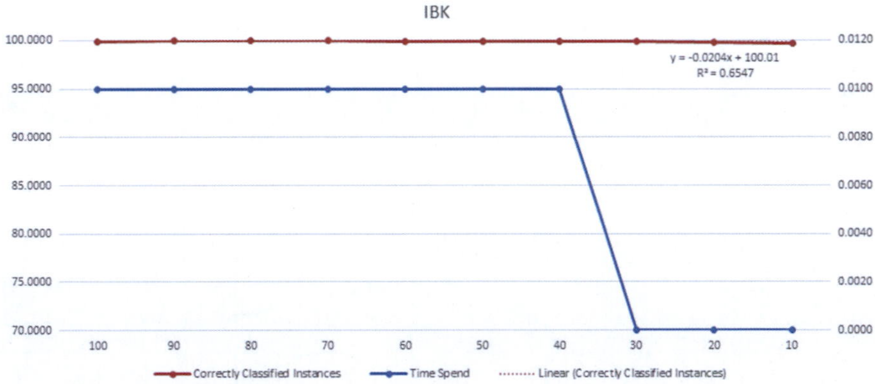

Fig. 6 Result for IBK algorithm based on incremental classification method

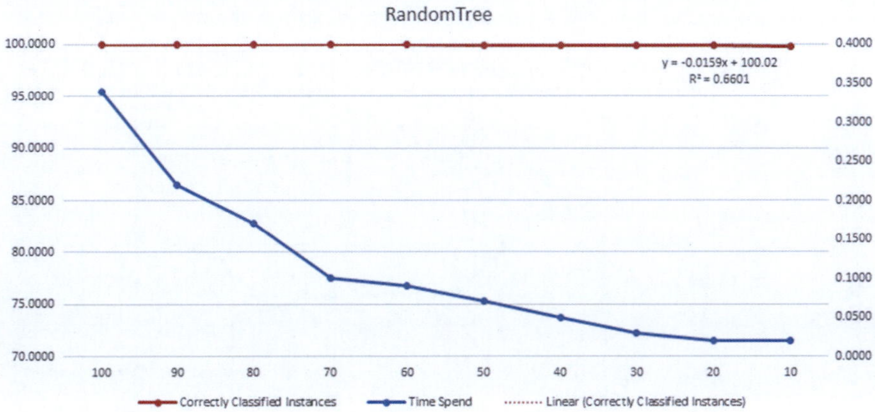

Fig. 7 Result for random tree algorithm based on incremental classification method

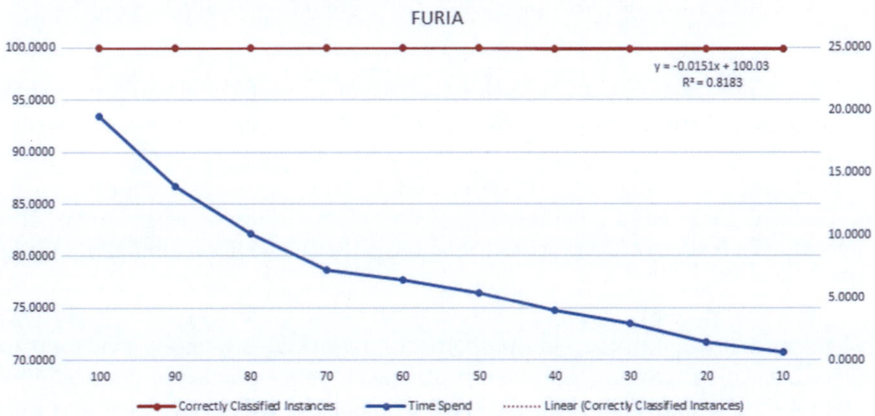

Fig. 8 Result for FURIA algorithm based on incremental classification method

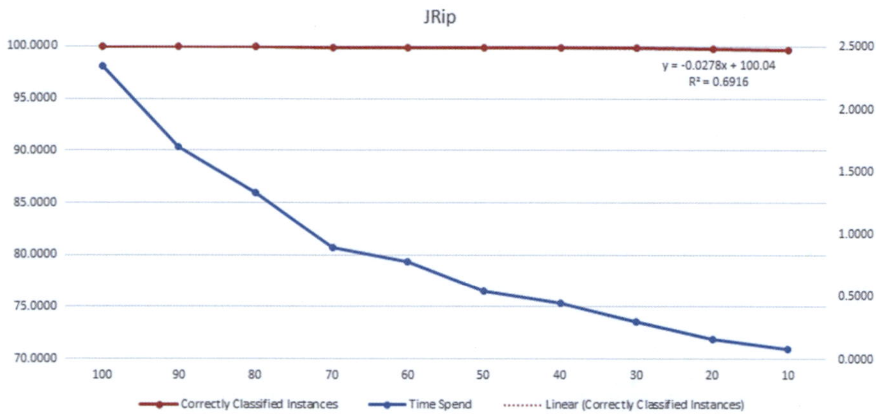

Fig. 9 Result for JRip algorithm based on incremental classification method

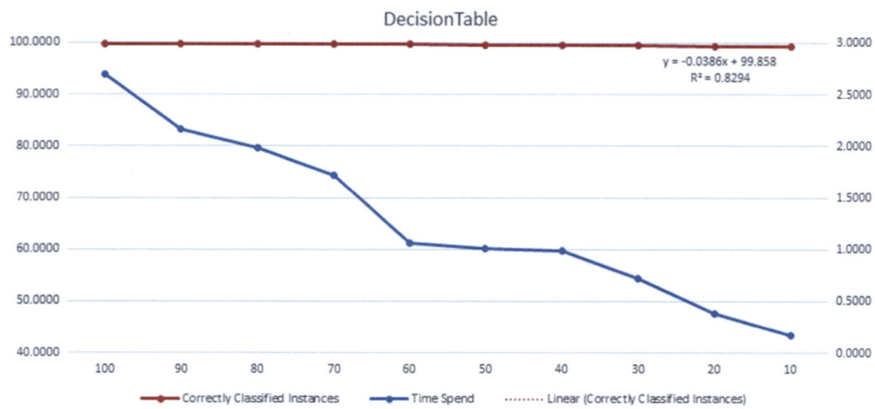

Fig. 10 Result for decision table algorithm based on incremental classification method

small sampling rate is used. The time spending curves for the remaining algorithms, random tree, FURIA, JRip, decision table, J48, follow a more or less exponential decline. Especially for random tree, although its time spending declines from 0.34 to 0.02 s, it is still efficient for detecting outliers in reality.

Decision table costs relatively longer time than JRip and J48. Even so, these algorithms just need 0.17 s to achieve 99% accuracy rate with sampling rate reaching 10%.

Algorithms in unstable group commonly have characteristics of drop in accuracy, when the sampling rates in a low proportion. "In some algorithms of this group, the accuracies fall to as low as 78% when insufficient training samples are present." Even when the "full dataset is made available for training up the models, their maximum accuracy ranges only from 79 to 98%." The "unstable curves of accuracy and the maximum accuracy, which is below 100% by certain extent, make these algorithms

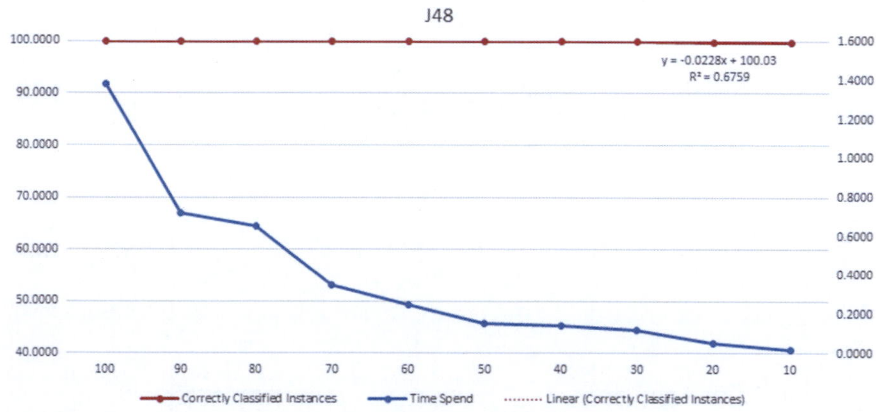

Fig. 11 Result for J48 algorithm based on incremental classification method

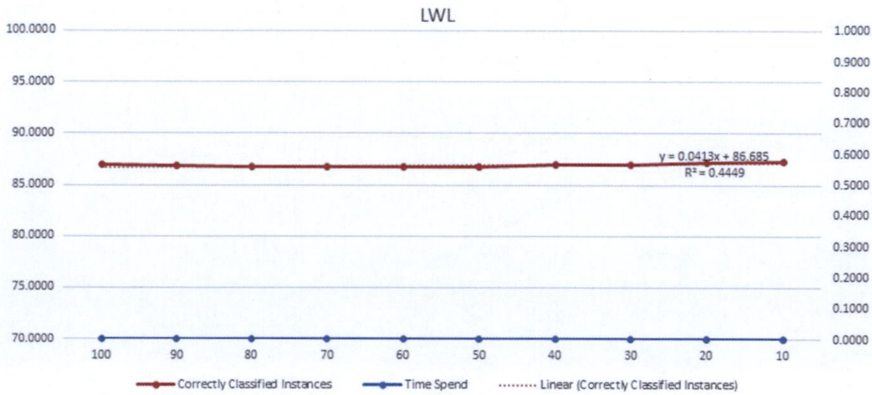

Fig. 12 Result for LWL algorithm based on incremental classification method

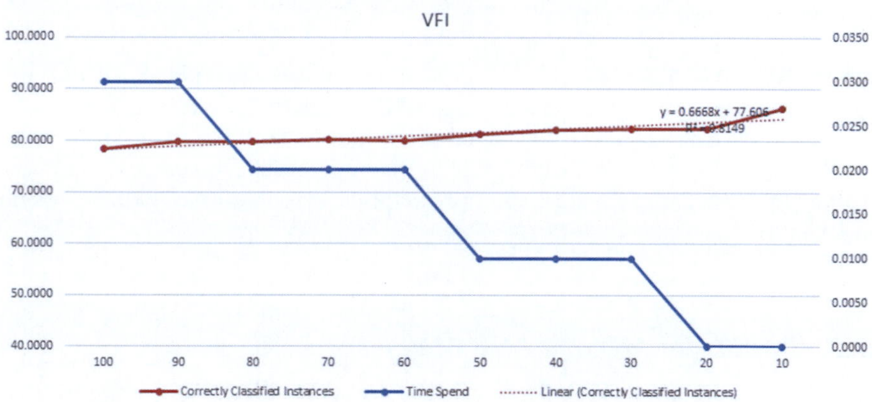

Fig. 13 Result for VFI algorithm based on incremental classification method

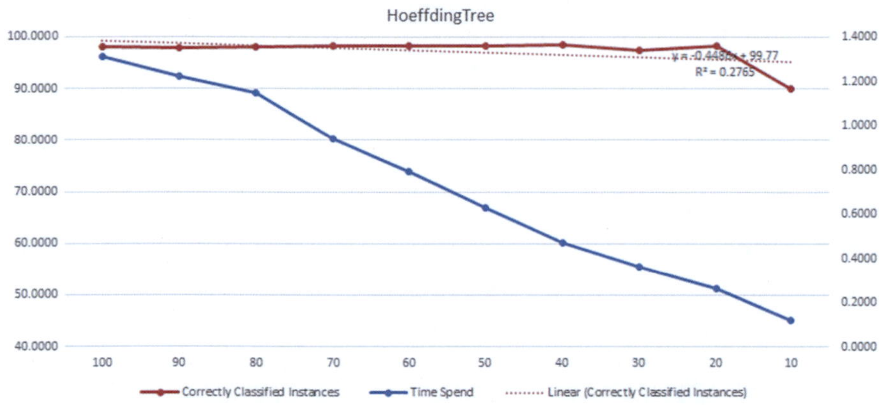

Fig. 14 Result for HoeffdingTree algorithm based on incremental classification method

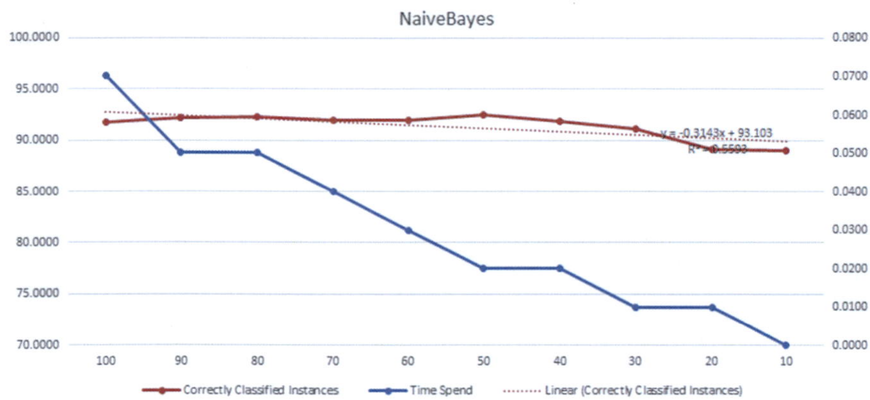

Fig. 15 Result for Naive Bayes algorithm based on incremental classification method

of this group a less favorable choice for incremental data mining that demands for a steady accuracy performance and quick model induction."

Due to the experiment we conducted above, it is better for us to the "incremental method to do the outlier detection in data streams." What is more, those algorithms belong to stable group are more appropriate for the classifier-based outlier detection method.

3.2 *Results for MD and LOF Methods*

Due to the incremental approach we applied in MD and LOF, we could get the plot for "number if outliers found" along with the changes of "number of instances processed.” In most cases, the one lays on the top position gains the best accuracy (Figs. 16, 17 and 18).

Table 1 shows the results from WEKA especially the classification function. A total of 10 classification algorithms are used for getting the time spend, accuracy, and other indicator variables like ROC area. We list the results in 3 separated columns; they are “Incre-100,” “m_l-0.5” and comparison. “Incre-100” stands for the results we get from the simple incremental method with full data, which we have discussed in Sect. 3.2 in this study. “m_l-0.5” stands for the best results we get from the MD and LOF method, which we have discussed earlier in this chapter. Here, the second indicator variable is lightweight analysis with medium sliding windows using Mahalanobis distance. The last column “comparison” obtained from the difference between these two.

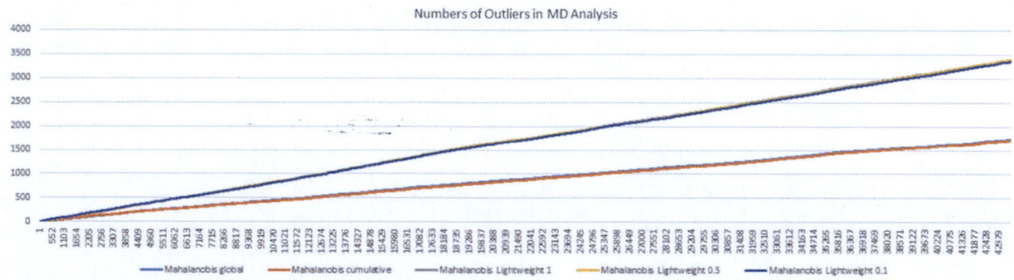

Fig. 16 Numbers of outliers in MD Analysis

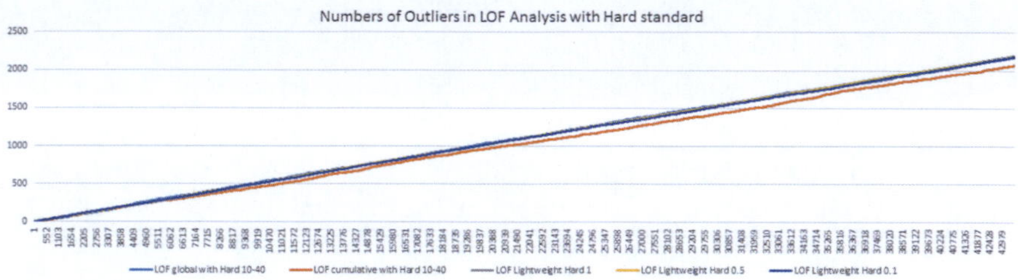

Fig. 17 Numbers of outliers in LOF analysis with hard standard

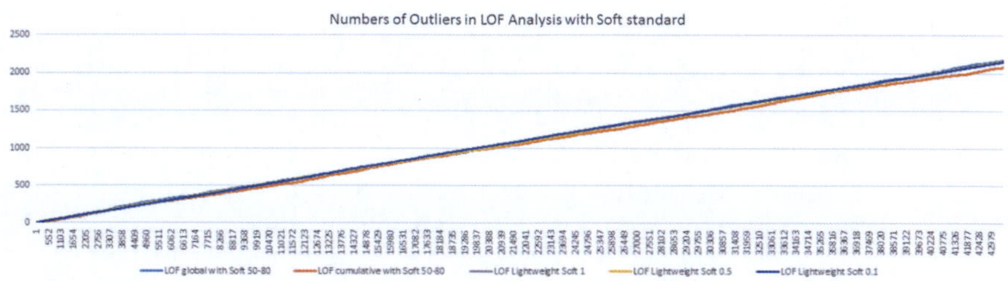

Fig. 18 Numbers of outliers in LOF analysis with soft standard

Table 1 Comparison among different outlier detection modes

Method	Dataset spilt Percent	Incre-100	m_l-0.5	Comparison
IBK	Time spend	0.01	0.01	0
IBK	Correctly classified instances	99.9278	99.9776	0.0498
IBK	Kappa statistic	0.998	0.9992	0.0012
IBK	Mean absolute error	0.0002	0.0001	−0.0001
IBK	Root-mean-squared error	0.0143	0.008	−0.0063
IBK	TP rate (weighted avg.)	0.999	1	0.001
IBK	FP rate (weighted avg.)	0.02	0	−0.02
IBK	Precision (weighted avg.)	0.999	1	0.001
IBK	Recall (weighted avg.)	0.999	1	0.001
IBK	F-measure (weighted avg.)	0.999	1	0.001
IBK	MCC (weighted avg.)	0.998	0.999	0.001
IBK	ROC area (weighted avg.)	0.999	1	0.001
VFI	Time spend	0.03	0.01	−0.02
VFI	Correctly classified instances	78.3696	91.0918	12.7222
VFI	Kappa statistic	0.5802	0.7535	0.1733
VFI	Mean absolute error	0.203	0.1734	−0.0296
VFI	Root-mean-squared error	0.2955	0.2637	−0.0318
VFI	TP rate (weighted avg.)	0.784	0.911	0.127
VFI	FP rate (weighted avg.)	0.005	0.002	−0.003
VFI	Precision (weighted avg.)	0.972	0.984	0.012
VFI	Recall (weighted avg.)	0.784	0.911	0.127
VFI	F-measure (weighted avg.)	0.861	0.945	0.084
VFI	MCC (weighted avg.)	0.675	0.802	0.127
VFI	ROC area (weighted avg.)	0.94	0.981	0.041
K-Star	Time spend	0	0	0
K-Star	Correctly classified instances	99.8828	99.9776	0.0948
K-Star	Kappa statistic	0.9967	0.9992	0.0025
K-Star	Mean absolute error	0.0006	0.0003	−0.0003
K-Star	Root-mean-squared error	0.0164	0.0081	−0.0083
K-Star	TP rate (weighted avg.)	0.999	1	0.001
K-Star	FP rate (weighted avg.)	0.003	0.001	−0.002
K-Star	Precision (weighted avg.)	0.999	1	0.001
K-Star	Recall (weighted avg.)	0.999	1	0.001
K-Star	F-measure (weighted avg.)	0.999	1	0.001
K-Star	MCC (weighted avg.)	0.997	0.999	0.002
K-Star	ROC area (weighted avg.)	1	1	0

(continued)

Table 1 (continued)

Method	Dataset spilt Percent	Incre-100	m_l-0.5	Comparison
Deci.T	Time spend	2.69	2.99	0.3
Deci.T	Correctly classified instances	99.7356	99.9202	0.1846
Deci.T	Kappa statistic	0.9926	0.9973	0.0047
Deci.T	Mean absolute error	0.0052	0.0037	−0.0015
Deci.T	Root-mean-squared error	0.031	0.0192	−0.0118
Deci.T	TP rate (weighted avg.)	0.997	0.999	0.002
Deci.T	FP rate (weighted avg.)	0.008	4	3.992
Deci.T	Precision (weighted avg.)	0.997	0.999	0.002
Deci.T	Recall (weighted avg.)	0.997	0.999	0.002
Deci.T	F-measure (weighted avg.)	0.997	0.999	0.002
Deci.T	MCC (weighted avg.)	0.993	0.997	0.004
Deci.T	ROC area (weighted avg.)	0.998	0.999	0.001
FURIA	Time spend	19.5	4.9	-14.6
FURIA	Correctly classified instances	99.977	99.985	0.008
FURIA	Kappa statistic	0.9994	0.9995	0.0001
FURIA	Mean absolute error	0.0001	0.0001	0
FURIA	Root-mean-squared error	0.0073	0.006	−0.0013
FURIA	TP rate (weighted avg.)	1	1	0
FURIA	FP rate (weighted avg.)	0	0	0
FURIA	Precision (weighted avg.)	1	1	0
FURIA	Recall (weighted avg.)	1	1	0
FURIA	F-measure (weighted avg.)	1	1	0
FURIA	MCC (weighted avg.)	1	1	0
FURIA	ROC area (weighted avg.)	1	1	0
JRip	Time spend	2.34	1.17	−1.17
JRip	Correctly classified instances	99.9586	99.9776	0.019
JRip	Kappa statistic	0.9988	0.9992	0.0004
JRip	Mean absolute error	0.0002	0.0001	−0.0001
JRip	Root-mean-squared error	0.0106	0.0078	−0.0028
JRip	TP rate (weighted avg.)	1	0.1	−0.9
JRip	FP rate (weighted avg.)	0	0	0
JRip	Precision (weighted avg.)	1	1	0
JRip	Recall (weighted avg.)	1	1	0
JRip	F-measure (weighted avg.)	1	1	0
JRip	MCC (weighted avg.)	0.999	0.999	0
JRip	ROC area (weighted avg.)	1	1	0

(continued)

Table 1 (continued)

Method	Dataset spilt Percent	Incre-100	m_l-0.5	Comparison
J48	Time spend	1.38	0.42	−0.96
J48	Correctly classified instances	99.9609	99.9626	0.0017
J48	Kappa statistic	0.9989	0.9987	−0.0002
J48	Mean absolute error	0.0002	0.0001	−0.0001
J48	Root-mean-squared error	0.0105	0.0101	−0.0004
J48	TP rate (weighted avg.)	1	1	0
J48	FP rate (weighted avg.)	0.001	0.001	0
J48	Precision (weighted avg.)	1	1	0
J48	Recall (weighted avg.)	1	1	0
J48	F-measure (weighted avg.)	1	1	0
J48	MCC (weighted avg.)	0.999	0.999	0
J48	ROC area (weighted avg.)	1	0.999	−0.001
Naive.B	Time spend	0.07	0.16	0.09
Naive.B	Correctly classified instances	91.7446	93.6007	1.8561
Naive.B	Kappa statistic	0.7562	0.756	−0.0002
Naive.B	Mean absolute error	0.0289	0.0186	−0.0103
Naive.B	Root-mean-squared error	0.1319	0.1108	−0.0211
Naive.B	TP rate (weighted avg.)	0.917	0.936	0.019
Naive.B	FP rate (weighted avg.)	0.21	0.264	0.054
Naive.B	Precision (weighted avg.)	0.941	0.947	0.006
Naive.B	Recall (weighted avg.)	0.917	0.936	0.019
Naive.B	F-measure (weighted avg.)	0.921	0.935	0.014
Naive.B	MCC (weighted avg.)	0.767	0.769	0.002
Naive.B	ROC area (weighted avg.)	0.975	0.988	0.013
LWL	Time spend	0	0	0
LWL	Correctly classified instances	86.9376	90.6379	3.7003
LWL	Kappa statistic	0.6672	0.7218	0.0546
LWL	Mean absolute error	0.0473	0.0321	−0.0152
LWL	Root-mean-squared error	0.1517	0.1242	−0.0275
LWL	TP rate (weighted avg.)	0.869	0.906	0.037
LWL	FP rate (weighted avg.)	0.037	0.022	−0.015
LWL	Precision (weighted avg.)	0.865	0.919	0.054
LWL	Recall (weighted avg.)	0.869	0.906	0.037
LWL	F-measure (weighted avg.)	0.856	0.905	0.049
LWL	MCC (weighted avg.)	0.747	0.774	0.027
LWL	ROC area (weighted avg.)	0.994	0.997	0.003

(continued)

Table 1 (continued)

Method	Dataset spilt Percent	Incre-100	m_l-0.5	Comparison
Rand.T	Time spend	0.34	0.18	−0.16
Rand.T	Correctly classified instances	99.9586	99.985	0.0264
Rand.T	Kappa statistic	0.9988	0.9995	0.0007
Rand.T	Mean absolute error	0.0001	0	−0.0001
Rand.T	Root-mean-squared error	0.0109	0.0065	−0.0044
Rand.T	TP rate (weighted avg.)	1	1	0
Rand.T	FP rate (weighted avg.)	0	0	0
Rand.T	Precision (weighted avg.)	1	1	0
Rand.T	Recall (weighted avg.)	1	1	0
Rand.T	F-measure (weighted avg.)	1	1	0
Rand.T	MCC (weighted avg.)	0.999	1	0.001
Rand.T	ROC area (weighted avg.)	1	1	0
Hoeff.T	Time spend	1.31	1.7	0.39
Hoeff.T	Correctly classified instances	98.1793	99.601	1.4217
Hoeff.T	Kappa statistic	0.948	0.9862	0.0382
Hoeff.T	Mean absolute error	0.0061	0.0018	−0.0043
Hoeff.T	Root-mean-squared error	0.0696	0.0307	−0.0389
Hoeff.T	TP rate (weighted avg.)	0.982	0.996	0.014
Hoeff.T	FP rate (weighted avg.)	0.053	0.017	−0.036
Hoeff.T	Precision (weighted avg.)	0.983	0.996	0.013
Hoeff.T	Recall (weighted avg.)	0.982	0.996	0.014
Hoeff.T	F-measure (weighted avg.)	0.982	0.996	0.014
Hoeff.T	MCC (weighted avg.)	0.951	0.986	0.035
Hoeff.T	ROC area (weighted avg.)	0.993	0.998	0.005

In most cases, we find the average time increased, but they changed in a level of hundred milliseconds. Time consuming to bring the lift on the correct rate, this is what we want to see. Correctly classified instances, another name for "accuracy," have improved generally. No matter 0.04% for random tree or 13% or VFI, accuracy really improved.

3.3 *Results for COD Methods*

As the key step aforesaid, we put the whole data to get the IQR value in process tab. In the directory: Filter ⋙unsupervised⋙attribute, we find "Interquartile Range, a filter for detecting outliers and extreme values based on interquartile ranges." IQR is "unsupervised" because it does not need a class label as a prerequisite. IQR belongs

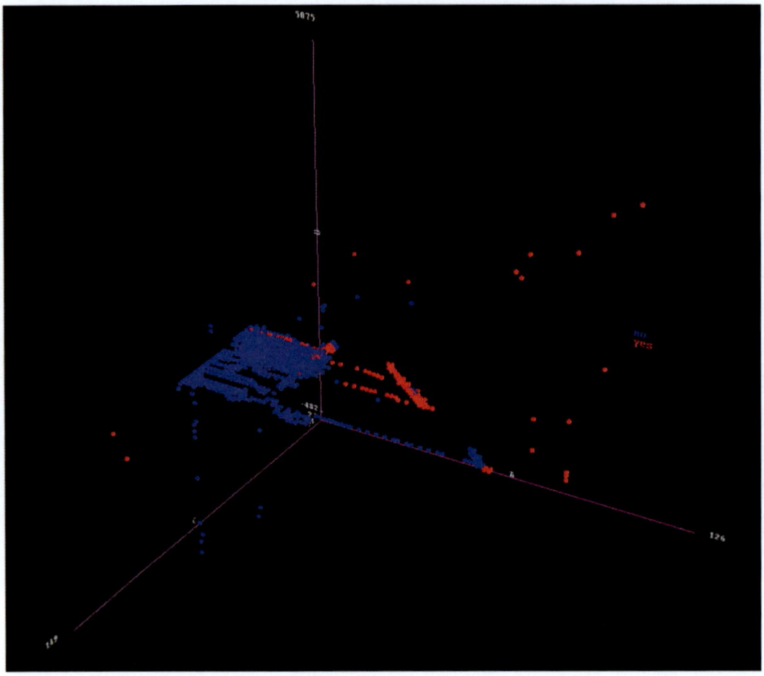

Fig. 19 Outlier distribution for global with IQR

to "attribute" is matter it generates the new column. From Fig. 19, we could easily find the instance is an outlier or not. The red one is outlier, while the blue one not. The user interface shows as follows.

In cumulative or lightweight, we also need a small set of data as a test data. Only we get the test data, we could do the train part. Cumulative method requires us to randomly get the test dataset. Here, we define 1000 instances for the test part and reservoir sampling for randomly selection. The outlier distribution for the test part is shown in Fig. 20.

Lightweight method also requires us to get the test dataset. Here, we define the first 1000 instances for the test part. The outlier distribution for the test part is shown in Fig. 21.

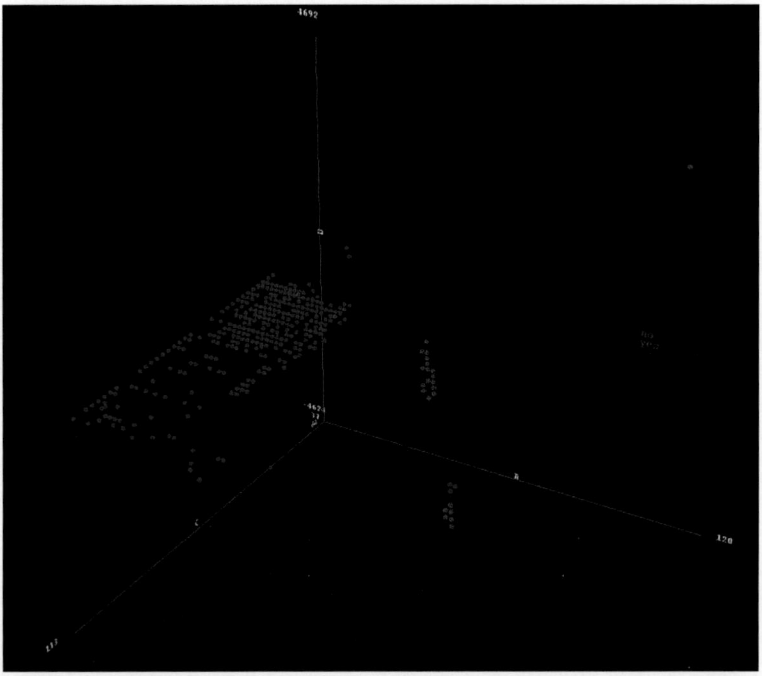

Fig. 20 Outlier distribution for random instances with IQR

As above results show, each time, we do the classification after detection. We call it classifier-based outlier detection, or COD in short. Only we do classification algorithms, we will get the "accuracy" to prove which outlier detection method is better.

Table 2 shows the outliers detection results comparing Mahalanobis and LOF with IQR. We calculate IQR value for each instance in the dataset to detect the outliers. Here, "global" means the whole dataset; then, we treat this "global" as a reference. We assume the "Hit Rate" for "Global_IQR" itself is 100%. "Total Outliers Hit" is the intersection of the number of elements between "Global_IQR" and applied method. "Outliers" is the total number of outliers for applied method. "Time" means the time consuming for finding outliers (Fig. 22).

It is obvious to find that the time cost nearly 0 when using IQR to find the outlier, but other methods to find outliers always need a very long time for computation in MATLAB. Here, we use 0 and symbol "∞" to indicate the great magnitude contrast in time. When conducting the experiments on MATLAB, we always wait at least 2 h for MD, even more for LOF. The approximate time consuming in outlier detection for MD, LOF, and IQR shows in Table 3. However for IQR, the program finishes computing in a flash. Hence, the IQR preprocessing method is very significant and more suitable for high-speed data stream.

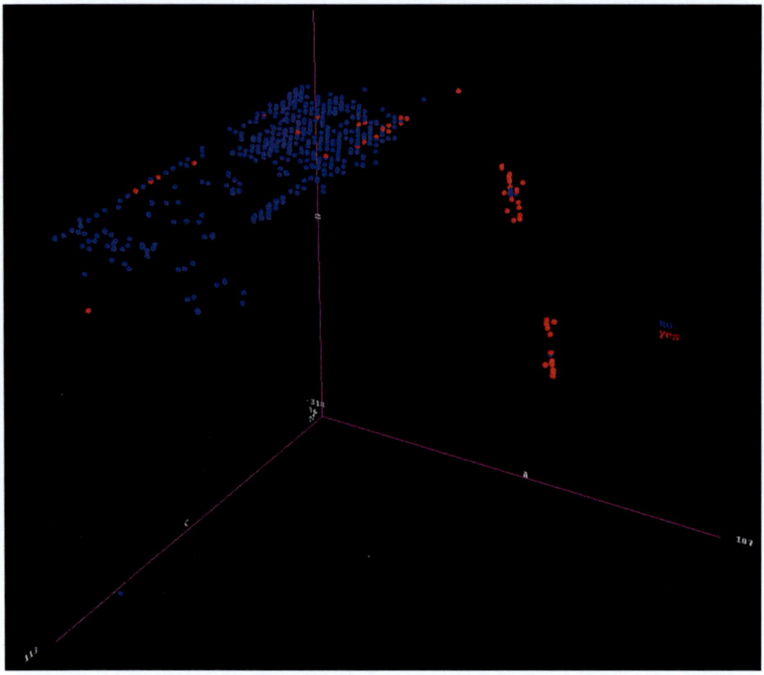

Fig. 21 Outlier distribution for first 1000 instances with IQR

This IQR is the basic calculation for COD. From step 1 of the experiment, we know how to select the classifier. From the step 2 of the experiments, we know which statistical method with analysis way gets the best accuracy, in regard to global analysis, cumulative analysis, or lightweight with sliding windows analysis. Here, our step 3 of the experiment calculates IQR value at first and then conducts the calculation for global analysis, cumulative analysis, or lightweight with sliding windows analysis, respectively. For the last step, we compare the result between result 2 and 3, under the classification algorithms from step 1. Although we compare Naive Bayes and VFI from unstable group, they are treated as the control group. The experimental group is resulted from classifiers IBK, JRip, decision table, and J48.

From the result shown in Table 4, the result using COD with different classifier is divided into two parts. The left part is the result after giving each instance an IQR value. The right part is the result from the best optimal parameters for MD or LOF. Obviously, we could set many parameters to calculate the MD or LOF value, but here the "Time" and "accuracy" are just the best one. Global analysis using LOF with hard standard, cumulative analysis using LOF with soft standard, and lightweight analysis with medium sliding windows using Mahalanobis distance have the best result for 3 kinds of analysis method in respective.

Firstly, we could find out most of the classification algorithms performing better in accuracy. Just a few of the classification results have 0.01–0.01 level's growth

Table 2 The hit rate compared to global IQR

43,500	Sum outliers hit	Hit rate	Outliers	Normal rate	Time find outliers
Global IQR	3471	100.000	3417	92.145	0
Mahalanobis global	868	25.007	1738	96.005	∞
Mahalanobis cumulative	1689	48.660	1719	96.048	∞
Mahalanobis lightweight 1	1437	41.400	3397	92.191	∞
Mahalanobis lightweight 0.5	1121	32.296	3401	92.182	∞
Mahalanobis lightweight 0.1	1171	33.737	3364	92.267	∞
LOF global with hard 10–40	277	7.980	2175	95.000	∞
LOF global with soft 50–80	107	3.083	2175	95.000	∞
LOF cumulative with hard 10–40	971	27.975	2074	95.232	∞
Cumulative with soft 50–80	818	23.567	2086	95.205	∞
LOF lightweight hard 1	1051	30.279	2200	94.943	∞
LOF lightweight soft 1	777	22.385	2160	95.034	∞
LOF lightweight hard 0.5	1031	29.703	2186	94.975	∞
LOF lightweight soft 0.5	786	22.645	2160	95.034	∞
LOF lightweight hard 0.1	1034	29.790	2186	94.975	∞
LOF lightweight soft 0.1	768	22.126	2160	95.034	∞

in accuracy percentage. Secondly, for the time aspect, although some of the time increased, it only takes 0.5 s more in building the classification model. We can conclude that the time spending applied COD method has no significant changes or bad impact.

Moreover, we find the accuracy of VFI after combining with COD suddenly drops. This is because VFI considers each feature separately, that is, a vector of feature values and a label for the class of the example. All classes vote for the distribution, and the sum of individuals vote forms the final vote of a class. This separately processing pattern is inconsistent with our attribute-related time-series data.

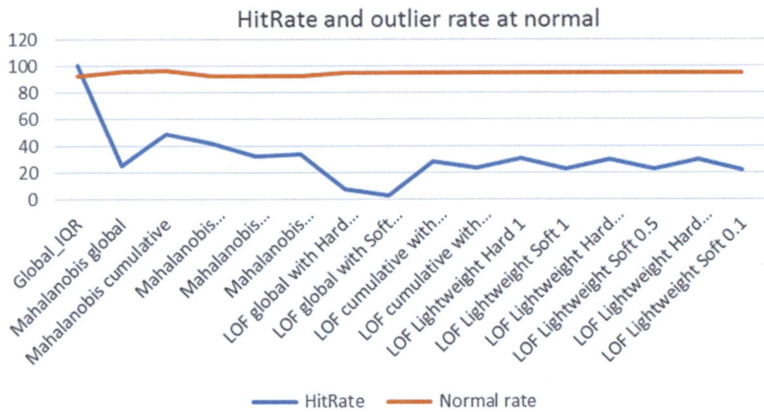

Fig. 22 Hit Rate and outlier rate at normal

Table 3 Approximate time consuming in outlier detection for MD, LOF, and IQR

Unit: second(s)	Global	Cumulative	Lightweight
Mahalanobis (MD)	1.00E+04	1.00E+05	1.00E+04
Local outlier factor (LOF)	1.00 +05	1.00E+06	1.00E+05
Interquartile range (IQR)	0	0	0

Table 4 The results for time and accuracy using COD based on different classifiers

Global					
	Accuracy	Time	Time	Accuracy	
Global_IQR_IBK	99.7573	0.01	0.01	99.9758	Global_IBK
Global_IQR_DecisionTable	99.8595	2.04	2.54	99.8621	Global_DecisionTable
Global_IQR_Jrip	99.9808	1.73	2.02	99.9782	Global_Jrip
Global_IQR_J48	99.9745	0.55	0.57	99.9782	Global_J48
Global_IQR_VFI	11.9668	0.02	0.02	92.9266	Global_VFI
Global_IQR_NaiveBayes	97.7075	0.05	0.05	93.0089	Global_NaiveBayes
Cumulative					
	Accuracy	Time	Time	Accuracy	
Cumulative_COD_IBK	99.8	0	0.01	99.9445	Cumulative_IBK
Cumulative_COD_DecisionTable	99.6	2.2	2.57	99.7996	Cumulative_DecisionTable
Cumulative_COD_Jrip	100	0.55	1.69	99.9662	Cumulative_Jrip
Cumulative_COD_J48	100	0	0.47	99.971	Cumulative_J48
Cumulative_COD_VFI	94.9	0	0.02	94.3858	Cumulative_VFI
Cumulative_COD_NaiveBayes	94.9	0	0.05	93.7169	Cumulative_NaiveBayes
Lightweight					
	Accuracy	Time	Time	Accuracy	
LightWeight_COD_IBK	96.5966	0.01	0.01	99.9776	LightWeight_IBK
LightWeight_COD_DecisionTable	96.3964	2.37	2.99	99.9202	LightWeight_DecisionTable
LightWeight_COD_Jrip	96.5966	1.77	1.17	99.9776	LightWeight_Jrip
LightWeight_COD_J48	96.4965	0.63	0.42	99.9626	LightWeight_J48
LightWeight_COD_VFI	12.9129	0.02	0.01	91.0918	LightWeight_VFI
LightWeight_COD_NaiveBayes	94.8949	0.06	0.16	93.6007	LightWeight_NaiveBayes

4 Performance Comparison in Root-Mean-Squared Error

The RMSE is a "quadratic scoring rule which measures the average magnitude of the error. The equation for the RMSE is given in both of the references." Expressing

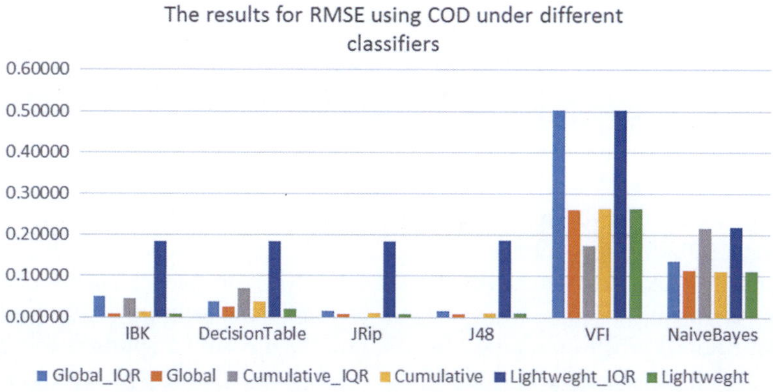

Fig. 23 Root-mean-squared error using COD under different classifiers

the formula in words, the "difference between forecast and corresponding observed values is each squared and then averaged over the sample." Finally, the "square root of the average is taken." Since the "errors are squared before they are averaged," the RMSE gives a "relatively high weight to large errors" which indicates that the RMSE is most "useful when large errors are particularly undesirable."

To simplify, we assume that there are n samples of model errors calculated as (e_i, $i = 1, 2, 3, \ldots, n$). The uncertainties brought in by "observation errors or the method used to compare model and observations are not considered here." We also assume "the error sample set is unbiased." The RMSE is calculated for the dataset as follows.

$$\text{RMSE} = \sqrt{\frac{1}{n}\sum_{i=1}^{n} e_i^2} \quad (8)$$

The underlying assumption when presenting the RMSE is that the errors are unbiased and follow a normal distribution (Fig. 23).

5 Summary and Research Directions

This chapter proposed a general framework to find outliers incrementally. Different methods on operation modes and distance measurements are expressed and experimented.

During the process that we are finding out the outliers, we should apply an "algorithm that is suitable for our dataset in terms of the correct distribution model, the correct attribute types," the number of instances, the running speed, "any incremental capabilities to allow new exemplars to be stored and the result accuracy."

Based on those attention factors, we choose to calculate the MD and LOF value of each instance. Combining different statistical analysis methods, we find the best accuracy result appeared in LOF with lightweight method at soft standard. However, if we put the preprocessing time into account, our classifier-based outlier detection method (COD), which calculates IQR as an evaluation variable and combines with classifiers, is better than any other outlier detection methods.

There would be two aspects to conduct in the future work aiming for the better openness, stability, and simplicity.

From the obvious aspect, like the up side of a coin, our results from experiments vary from different windows size and different measure standard. Hence, COD method is still facing a problem on how to choose the appropriate variables and get the highest accuracy.

From the implied aspect, like the bottom side of a coin, we move away those outliers in our experiments because they are unnecessary. What if we apply our proposed method on some rare disease detection cases? In this situation, outliers are much important than inliers.

Key Terminology and Definitions [each keyword to be explained in 5–10 sentences]

Data mining—An interdisciplinary subfield of computer science and is the computational process of discovering patterns in large datasets involving methods at the intersection of artificial intelligence, machine learning, statistics, and database systems.

Outlier—An observation point that is distant from other observations.

Algorithm—A self-contained step-by-step set of operations to be performed. Algorithms exist that perform calculation, data processing, and automated reasoning.

Dr. Simon Fong graduated from La Trobe University, Australia, with a 1st Class Honors B.E. Computer Systems degree and a Ph.D. Computer Science degree in 1993 and 1998, respectively. He is now working as an Associate Professor at the Computer and Information Science Department of the University of Macau. He is a co-founder of the Data Analytics and Collaborative Computing Research Group in the Faculty of Science and Technology. Prior to his academic career, he took up various managerial and technical posts, such as systems engineer, IT consultant, and e-commerce director in Australia and Asia. He has published over 432 international conference and peer-reviewed journal papers, mostly in the areas of data mining, data stream mining, big data analytics, meta-heuristics optimization algorithms, and their applications. He serves on the editorial boards of the Journal of Network and Computer Applications of Elsevier (I.F. 3.5), IEEE IT Professional Magazine, (I.F. 1.661) and various special issues of SCIE-indexed journals. He is also an active researcher with leading positions such as Vice-chair of IEEE Computational Intelligence Society (CIS) Task Force on "Business Intelligence and Knowledge Management," and Vice-director of International Consortium for Optimization and Modelling in Science and Industry (iCOMSI).

Ms. Tengyue Li is currently a M.Sc student major in E-Commerce Technology at the Department of Computer and Information Science, University of Macau, Macau SAR of China. She participated in the university the following activities: Advanced Individual in the School, Second Prize in the Smart City APP Design Competition of Macau; Top 10 in the China Banking Cup Million

Venture Contest; Campus Ambassador of Love. She has internship experiences as a Meituan Technology Company Product Manager from June to August 2017. She worked at Training Base of Huawei Technologies Co., Ltd. from September to October 2016. From February to June 2016, she worked at Beijing Yangguang Shengda Network Communications as data analyst. Lately, she involved in projects such as "A Minutes" Unmanned Supermarket by the University of Macau Incubation Venture Project since September 2017.

Mr. Han Dong received the B.S. degree in electronic information science and technology from Beijing Information Science and Technology University (BISTU), China. He is currently pursuing his master degree in E-commerce technology in University of Macau, Macau S.A.R. of People's Republic of China. His current research focuses on the massive data analysis.

Dr. Sabah Mohammed research interest is in intelligent systems that have to operate in large, nondeterministic, cooperative, survivable, adaptive or partially known domains. Although his research is inspired by his PhD work back in 1981 (Brunel University, UK) on the employment of some brain activity structures-based techniques for decision making (planning and learning) that enable processes (e.g., agents, mobile objects) and collaborative processes to act intelligently in their environments to timely achieve the required goals. He is a full professor of Computer Science with Lakehead University, Ontario, Canada, since 2001 and Adjunct Research Professor with the University of Western Ontario since 2009. He is the Editor-in-Chief of the international journal of Ubiquitous Multimedia (IJMUE) since 2005. His research touches many areas including Web intelligence, big data, health informatics, and security of cloud-based EHRs among others.

Chapter 7
Comparison of Contemporary Meta-Heuristic Algorithms for Solving Economic Load Dispatch Problem

Simon Fong, Tengyue Li, and Zhiyan Qu

1 Introduction

The power system consists of many generates units and they consume the fuel to generate power. There exists power loss among different units during transmission. To solve the ELD is actually minimize the total fuel cost of all units considering the power loss. And the problem can be described mathematically in five formulas:

$$\text{Minimize} \sum_{i=1}^{n} \mathrm{F_i}(P_i) \tag{1}$$

$$F_{ij}(P_i) = a_i P_i^2 + b_i P_i + c_i, \quad P_i^{\min} \le P_i \le P_i^{\max} \tag{2}$$

where

P_i: Output power generation of unit *i*.

a_i, b_i, c_i: Fuel cost coefficients of unit *i*.

$$\sum_{i=1}^{n} P_i = D + P_l \tag{3}$$

D: Total real power is demand

P_l: Total power losses.

S. Fong (✉) · T. Li · Z. Qu
Department of Computer Science, University of Macau, Taipa, Macau SAR
e-mail: ccfong@umac.mo

T. Li
e-mail: litengyue2018@gmail.com

S. J. Fong and R. C. Millham (eds.), *Bio-inspired Algorithms for Data Streaming and Visualization, Big Data Management, and Fog Computing*, Springer Tracts in Nature-Inspired Computing,
https://doi.org/10.1007/978-981-15-6695-0_7

$$P_l = \sum_{i}^{n} \sum_{j}^{n} B_{ij} P_i P_j \tag{4}$$

B is a square matrix of transmission coefficients.

(1) is the objective function and (3, 4) are constraints. By using the penalty function method, we can get only one formula (5) to be the objective function.

Minimize

$$\sum (a_i P_i^2 + b_i P_i + c_i) + 1000 * \text{abs}\left(\sum_{i=1}^{n} P_i - D - \sum_{i=1}^{n} \sum_{j=1}^{n} B_{ij} P_i P_j\right) \tag{5}$$

If without considering the transmission from one generator to another generator, P_l will be ignored, the objective function will be:

Minimize

$$\sum (a_i P_i^2 + b_i P_i + c_i) + 1000 * \text{abs}\left(\sum_{i=1}^{n} P_i - D\right) \tag{6}$$

If the objective function is not got by the penalty method, then valve pointing is always considered (Sinha et al. 2003; Yang et al. 2012). The valve -point effects introduce ripples in the heat-curves because the generating units with multivalve stream turbines exhibit a greater variation in the fuel cost functions (Sinha et al. 2003). The objective function can be given:

Minimize

$$\sum (a_i P_i^2 + b_i P_i + c_i) + |e_j \times \sin(f_j \times (P_{i\,\min} - P_j))| \tag{7}$$

So our purpose turns into minimizing these objective functions. Because situation (6) is the simplified version of (5) and (6), it will not be tested by cases. (5) and (6) will be tested under two different cases, the details are in the following section.

In the later years, many efficient new heuristic algorithms are invented or improved algorithms perform efficiently. They are mainly developed by Xin-She Yang and being tried to solve many problems especially the N-P hard problem, multi-objective problem, etc. In this paper, these algorithms are used to solve the ELD problem and compared their results and efficiency. The new algorithm is Firefly Algorithm (FA) (Yang 2009), Cuckoo Search (CS) algorithm (Yang and Deb 2009), Bat Algorithm (BA) (Yang 2010a), Flower Pollination Algorithm (FPA) (Yang 2012) and MFA which is developed by Luo (Gao et al. 2015) and WSA (Tang et al. 2012). Because the Particle Swarming Optimization (PSO) algorithm (Kennedy and Eberhart 1995) is proved that it has the absolute advantage over Quadratic programming (QP) and GA in four different cases (Zaraki and Bin Othman 2009). QP is a traditional linear programming method the GA is a classical evolutionary programming method. Now that PSO is proved superior to most of the solutions for ELD problems, one thing is that there is no need to compare these new algorithms with the traditional methods such as QP and dynamic programming, another thing is that PSO can be used to be compared with the above algorithms as a benchmark to reflect these new algorithms' performance. Besides, in Ref. (Yang et al. 2012), FA is also proved a good method for ELD problem, as the latest algorithm, we can not only verify its efficiency and also compare it with other latest algorithms.

These different algorithms all having their unique characters. PSO is based on the swarming behavior of fish and birds and developed by Kennedy and Eberhart (1995). It consists of mainly mutation and selection and converges very quickly but may lead to premature convergence (Yang 2010b). FA was developed by Yang in 2008 and is based on the flashing behavior of swarming fireflies (Yang 2009). Attraction is firstly used and local attraction is stronger than long-distance attraction. It let the subgroup swarm around a local mode and can deal with the multimodal problems efficiently (Yang 2010b). MFA is developed from FA by Luo and used the greedy idea. The greedy idea is focusing on the individual who didn't reach the known best point, to analyze each of its coordinate parameters, and exchange it with the gained best firefly's coordinate parameters (Gao et al. 2015). It is more efficient than FA in high dimension problems and global optimization. CS is developed by Yang and Suash Deb in 2009 and is based on brooding parasitism of cuckoo and is enhanced by the so-called Levy-flight (2009). It has efficient random walks and balanced mixing and very efficient in global search (Yang 2010b). BA is developed by Yang in 2010 and is based on the echolocation of the forging bat (Yang 2010a). It firstly uses frequency tuning thus the mutation varies due to the variations of the bat loudness and pulse emission (Yang 2010b). FPA was developed by Yang in 2012 that is based on the flower pollination process. Flower pollination (mutation) activities can occur at all scales, both local and global. It has been extended to multi-objective problems. WSA is developed by Tang and Simon in 2012 and is based on wolf hunting and escaping behavior (Tang et al. 2012). It only has the local mutation and uses the jump probability to avoid being caught in the local modal. The summary of these algorithms is in Table 1 in the appendix.

Table 1 Wolf search algorithm (WSA) parameters

Parameter	Value	Description
popsize	25	The number of search agent (population)
Visual	1	Visual distance
pa	0.25	Escape possibility
coordinatesSize	10	The length of the coordinates
largestGeneration	10,000	The maximum generation allowed
Gamma0	1.0	
Alpha	1.0	Randomness 0–1

2 Experiment

The testing environment is RAM: 8 GB, CPU: 3.6GHZ 64 bit. In order to get the best performance of each algorithm, each of them run 50 times, then the best result, average, worst, standard deviation of the total fuel cost value are considered. They all use 25 agents and iterate 10,000 times. The algorithms' code is obtained from the web sharing files (http://www.mathworks.com/matlabcentral/fileexchange/7506-particle-swarm-optimization-toolbox; http://www.mathworks.com/matlabcentral/fileexchange/2969; http://www.mathworks.com/matlabcentral/fileexchange/37582-bat-algorithm--demo-; http://www.mathworks.com/matlabcentral/fileexchange/45112-flower-pollination-algorithm) except WSA and MFA.

The following Tables 1, 2, 3, 4, 5, and 6 are parameters of each algorithm.

Table 2 Firefly algorithm (FA) and maniac firefly algorithm (MFA) parameters

Parameter	Value	Description
n	25	The number of search agent (population)
MaxGeneration	10,000	Number of pseudo time steps
Alpha	0.5	Randomness 0–1 (highly random)
betamn	0.2	Minimum value of beta
Gamma	1	Absorption coefficient

Table 3 Flower pollination algorithm (FPA) parameters

Parameter	Value	Description
n	25	The number of search agent (population) (10–25)
p	0.2	Probability switch
largestGeneration	10,000	Total number of iterations

Table 4 Bat algorithm (BA) parameters

Parameter	Value	Description
n	25	The number of search agent (population) (10–40)
N_gen	10,000	Number of generations
A	0.5	Loudness (constant or decreasing)
r	0.5	Pulse rate (constant or decreasing)
$Q_{\min}$	0	Frequency minimum
$Q_{\max}$	2	Frequency minimum

Table 5 Cuckoo search (CS) parameters

Parameter	Value	Description
n	25	The number of search agent (population)
pa	0.25	Discovery rate of alien eggs/solutions
times	10,000	Number of iterations

Table 6 Particle search optimization (PSO) parameters

Parameter	Value	Description
df	100	Epochs between updating display
me	2000	Maximum number of iterations
ps	25	Population size
ac1	2	Acceleration const 1 (local best influence)
ac2	2	Acceleration const 2 (global best influence)
iw1	0.9	Initial inertia weight
iw2	0.4	Final inertia weight
iwe	1500	Epoch when inertial weight at final value
ergrd	1e−25	Minimum global error gradient
ergrdep	150	Epochs before error gradient criterion Terminates run
errgoal	NaN	Error goal
trelea	0	Type flag (which kind of PSO to use)
PSOseed	0	PSOseed

3 Testing Cases

There are totally four cases, the first and second use the objective function (5), and the third and fourth use the objective function (6). Each case is tested under seven algorithms and the cases are considered in turn from small scale 3 to the scale 40.

3.1 Case 1

This test case includes 3 generating units. The load demand is $P_d = 150$ MW. In this case, the penalty is method is used in the cost function and transmission loss is considered. The coefficients value refers to and shown (Table 7).

And the testing result is in Table 8.

Table 7 Fuel cost function coefficient of three generating units

Plant no.	a_i ($/MW2)	b_i ($/MW)	c_i ($)	$P_{\min}$ (Mw)	P_{max} (Mw)
1	0.008	7	200	10	85
2	0.009	6.3	180	10	80
3	0.007	6.8	140	10	70

$B = 0.01$ * [0.0218 0.0093 0.0028; 0.0093 0.0228 0.0017; 0.0028 0.0017 0.0179]

Table 8 Fuel cost function coefficient of six generating units

Plant no.	a_i	b_i	c_i	$P_{\min}$	$P_{\max}$
	($/MW2)	($/MW)	($)	(Mw)	(Mw)
1	0.007	7	240	100	500
2	0.0095	10	200	50	200
3	0.009	8.5	220	80	300
4	0.009	11	200	50	150
5	0.008	10.5	220	50	200
6	0.0075	12	120	50	120

$B = 1e{-}4$ * [0.14 0.17 0.15 0.19 0.26 0.22; 0.017 0.6 0.13 0.16 0.15 0.2; 0.015 0.13 0.65 0.17 0.24 0.19; 0.019 0.16 0.17 0.71 0.3 0.25; 0.026 0.15 0.24 0.3 0.69 0.32; 0.022 0.2 0.19 0.25 0.32 0.85]

3.2 *Case 2*

This test case includes 6 generating units and the scale became larger. The load demand is $P_d = 700$ MW. In this case, the penalty is method is also used in the cost function and transmission loss is considered. The coefficients value refers to (Saadat 1999) and shown below.

3.3 *Case 3*

This test case includes 13 generating units. In this large system, the load demand is $P_d = 700$ MW. Because it is a higher non-linear space, time needed more to seek for the solution. In this case, the valve pointing is considered and ignore the transmission loss. The coefficients value of the cost function refers to (Sinha et al. 2003) and shown below (Table 9).

Table 9 Fuel cost function coefficient of 13 generating units

Plant no	a_i	b_i	c_i	e_i	f_i	$P_{\min}$	$P_{\max}$
	($/MW2)	($/MW)	($)	($)	($)	(Mw)	(Mw)
1	0.00028	8.1	550	300	0.035	0	680
2	0.00056	8.1	309	200	0.042	0	360
3	0.00056	8.1	307	200	0.042	0	360
4	0.00324	7.74	240	150	0.063	60	180
5	0.00324	7.74	240	150	0.063	60	180
6	0.00324	7.74	240	150	0.063	60	180
7	0.00324	7.74	240	150	0.063	60	180
8	0.00324	7.74	240	150	0.063	60	180
9	0.00324	7.74	240	150	0.063	60	180
10	0.00284	8.6	126	100	0.084	40	120
11	0.00284	8.6	126	100	0.084	40	120
12	0.00284	8.6	126	100	0.084	55	120
13	0.00284	8.6	126	100	0.084	55	120

3.4 *Case 4*

This test case includes 40 generating units. The load demand is $P_d = 10{,}500$ MW. The data of this case is from (Chen and Chang 1995; Sinha et al. 2003). This case is used to test the limits of the algorithms to deal with ELD problem. Because its solution space is large enough and more local minima to trap the algorithms' agents. So this is a good case to test the algorithm exploration. The function coefficients value are as follows (Table 10).

3.5 *Testing Results and Analysis*

Now that our objective is to minimize the fuel cost on condition that the limits are satisfied, the fuel cost that is also the fitness of the algorithm is mainly listed and compared. They are listed in Tables 11, 12, 13, and 14. Also, the box plots are followed presenting the fuel cost values distribution in 50 different times. They are illustrated in Figs. 1, 2, 3, and 4.

In this case, the FPA and CS get the best fitness value which means get the lowest fuel cost and the best solution for ELD problem. And from the deviation and box plot, we can see these two algorithms perform well and steadily in this case. FA and MFA also perform well but not as well as the former two algorithms.

In this case, CS still performs best and FPA, FA, MFA followed. And CS and FPA still steady ones.

Table 10 Fuel cost function coefficient of 40 generating units

Plant no	a_i	b_i	c_i	e_i	f_i	$P_{\min}$	$P_{\max}$
	($/MW2)	($/MW)	($)	($)	($)	(Mw)	(Mw)
1	0.0069	6.73	94.705	100	0.084	36	114
2	0.0069	6.73	94.705	100	0.084	36	114
3	0.0203	7.07	309.54	100	0.084	60	120
4	0.0094	8.18	369.54	150	0.063	80	190
5	0.0114	5.35	148.89	120	0.077	47	97
6	0.0114	8.05	222.33	100	0.084	68	140
7	0.0036	8.03	287.71	200	0.042	110	300
8	0.0049	6.99	391.98	200	0.042	135	300
9	0.0057	6.6	455.76	200	0.042	135	300
10	0.0061	12.9	722.82	200	0.042	130	300
11	0.0052	12.9	635.2	200	0.042	94	375
12	0.0057	12.8	654.69	200	0.042	94	375
13	0.0042	12.5	913.4	300	0.035	125	500
14	0.0075	8.84	1760.4	300	0.035	125	500
15	0.0071	9.15	1728.3	300	0.035	125	500
16	0.0071	9.15	1728.3	300	0.035	125	500
17	0.0031	7.97	647.83	300	0.035	220	500
18	0.0031	7.97	647.83	300	0.035	220	500
19	0.0031	7.97	647.83	300	0.035	242	550
20	0.0031	7.97	647.83	300	0.035	242	550
21	0.003	6.63	785.96	300	0.035	254	550
22	0.003	6.63	785.96	300	0.035	254	550
23	0.0028	6.66	794.53	300	0.035	254	550
24	0.0028	6.66	79,453%	300	0.035	254	550
25	0.0028	7.1	801.32	300	0.035	254	550
26	0.0028	7.1	801.32	300	0.035	254	550
27	0.5212	3.33	1055.1	120	0.077	10	150
28	0.5212	3.33	1055.1	120	0.077	10	150
29	0.5212	3.33	1055.1	120	0.077	10	150
30	0.0114	5.35	148.89	120	0.077	47	97
31	0.0016	6.43	222.92	150	0.063	60	190
32	0.0016	6.43	222.92	150	0.063	60	190
33	0.0016	6.43	222.92	150	0.063	60	190
34	0.0001	8.62	116.58	200	0.042	90	200
35	0.0001	8.62	116.58	200	0.042	90	200

(continued)

Table 10 (continued)

Plant no	a_i ($/MW2)	b_i ($/MW)	c_i ($)	e_i ($)	f_i ($)	$P_{\min}$ (Mw)	$P_{\max}$ (Mw)
36	0.0001	8.62	116.58	200	0.042	90	200
37	0.0161	5.88	307.45	80	0.098	25	110
38	0.0161	5.88	307.45	80	0.098	25	110
39	0.0161	5.88	307.45	80	0.098	25	110
40	0.0031	7.97	647.83	300	0.035	242	550

Table 11 The fuel cost in case 1 with 3 units

Case 1	Fuel cost ($/h)			
Algorithm	Best	Average	Worst	Standard deviation
BA	1580.059891	1590.230917	1622.970969	10.697375
CS	**1579.928213**	**1579.928213**	**1579.928213**	**0.000000**
FA	1579.928242	1579.930793	1579.942414	0.003373
FPA	**1579.928214**	**1579.928216**	1579.928222	**0.000002**
MFA	1579.928242	1579.930793	1579.942414	0.003373
PSO	1579.967437	1580.619795	1583.919284	0.714496
WSA	1579.928648	1581.755419	1588.06562	2.035278

Table 12 The fuel cost in case 1 with 6 units

Case 2	Generation cost ($/h)			
Algorithm	Best	Average	Worst	Standard deviation
BA	8282.956323	8623.087705	9025.261096	201.119261
CS	**8229.377561**	**8229.377561**	**8229.377561**	**0.000000**
FA	**8229.378531**	8229.814447	8230.839047	0.381278
FPA	**8229.379183**	8229.382225	8229.391161	**0.002372**
MFA	**8229.378531**	8229.814447	8230.839047	0.381278
PSO	8510.013191	9237.097437	15374.1546	1355.662544
WSA	8244.86442	8312.229494	8417.335	37.917384

In this case, the solution is in a higher space, CS still perform well, so does FPA, but we can see that their advantage solution exploitation is degrading because BA, FA, MFA are all getting the same best fitness as CS.

In this case with high complexity, PSO is outstanding. Although it does not perform so steadily, which means it is easy to trap into a local minimum, its whole performance exceeds other algorithms.

From case 1 and case 4, we can conclude that:

Table 13 The fuel cost in case 1 with 13 units

Case 3	Generation cost ($/h)			
Algorithm	Best	Average	Worst	Standard deviation
BA	**7626.654000**	8199.430601	11852.04569	1110.031035
CS	**7626.654000**	**7626.654**	**7626.654000**	**3.67491E−12**
FA	**7626.654000**	7639.299648	7940.911835	62.17009879
FPA	**7626.654000**	**7626.654**	7626.654000	**3.67491E−12**
MFA	**7626.654000**	7633.345693	7958.060257	46.85917699
PSO	8627.379185	10216.63	12,028.29847	816.262326
WSA	8040.066032	9883.711072	12,390.98362	856.6355272

Table 14 The fuel cost in case 1 with 40 units

Case 4	Generation cost ($/h)			
Algorithm	Best	Average	Worst	Standard deviation
BA	148,922.272	163,071.1517	177,835.4155	7305.650424
CS	**143,746.2964**	143,751.9159	144,027.2739	**39.73621553**
FA	143,862.6583	144,965.4281	146,381.342	533.8345428
FPA	**143,746.2964**	143,999.9002	145,635.9741	422.3975895
MFA	143,920.2387	144,997.9737	146,402.9696	494.0217769
PSO	**129,397.7116**	**130,123.0574**	**130,708.9265**	**295.2216932**
WSA	151,364.8657	156,937.1785	161,362.7307	2397.841355

(a) BA performs worst no matter the best value it can seek for, or the steadiness of solution for the ELD problem.
(b) PSO performs not so good and steadily in the ELD problem. But for the complex ELD problems, PSO still has the possibility to seek for a better solution although it also tends to converge to a local optimum. So, for the very high dimension ELD problem (number of units $\geq$ 40), PSO is recommended.
(c) CS and FPA perform best in the low and medium ELD problems, especially CS. But they will meet their bottleneck in the very high dimension problem. So if the ELD is not too complex, CS is recommended.
(d) FA and MFA are also a good choice, they get nearly the same best results which is closer enough to CS from case 1 to case 4. The difference between FA and MFA can be got from their deviation result and also the box plots. All the cases show that MFA performs much more steadily than FA. So, MFA is a better choice compared to FA.
(e) we can see that from cases 1 to 4, WSA performs better than the BA, but it only performs well in the low dimension ELD problems which are cases 1 and 2.

Only considering the best situation of fuel cost results, the power needed to generate for each unit can refer to Table 15 in Appendix 1.

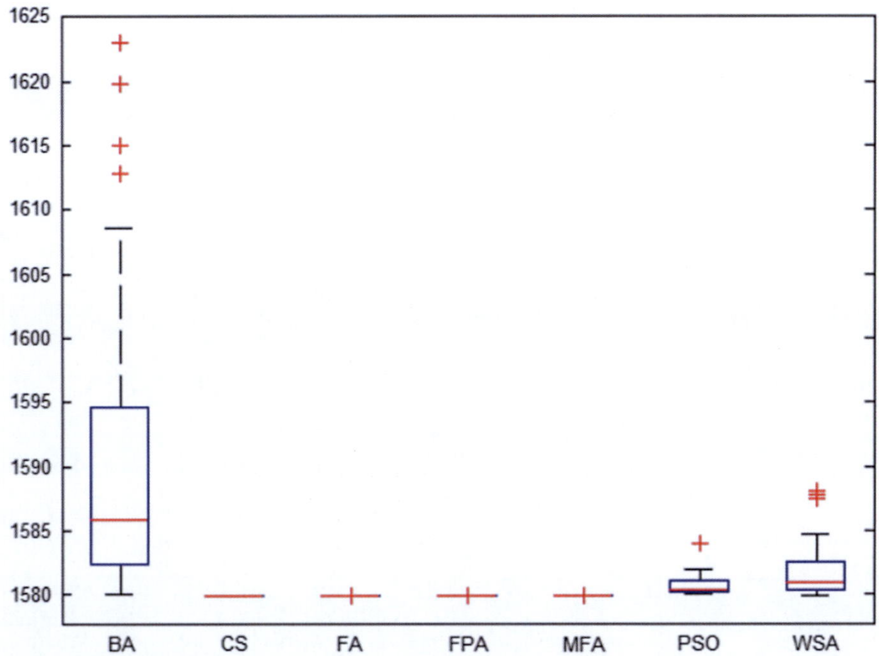

Fig. 1 The fuel cost in case 1 (50 times)

4 Conclusion

In this case study, we can see different algorithms performs differently in different cases. CS performs best except the high dimension (40 units) and so does FPA. Because CS combines local random walk and global exploration so well (Yang 2010b), it is able to perform so well and steadily. MFA adds the greedy idea to FA so that it can get more steady results than FA. WSA needs to be improved for its disadvantage in higher dimension problem. PSO is only temporarily recommended to the high dimension ELD problems.

Key Terminology and Definitions [each keyword to be explained in 5–10 sentences]

Metaheuristics—In computer science and mathematical optimization, a meta-heuristic is a higher-level procedure or heuristic designed to find, generate, or select a heuristic (partial search algorithm) that may provide a sufficiently good solution to an optimization problem, especially with incomplete or imperfect information or limited computation capacity (Chen and Chang 1995; Sinha et al. 2003). Metaheuristics sample a set of solutions which is too large to be completely sampled. Metaheuristics may make few assumptions about the optimization problem being solved, and so they may be usable for a variety of problems.

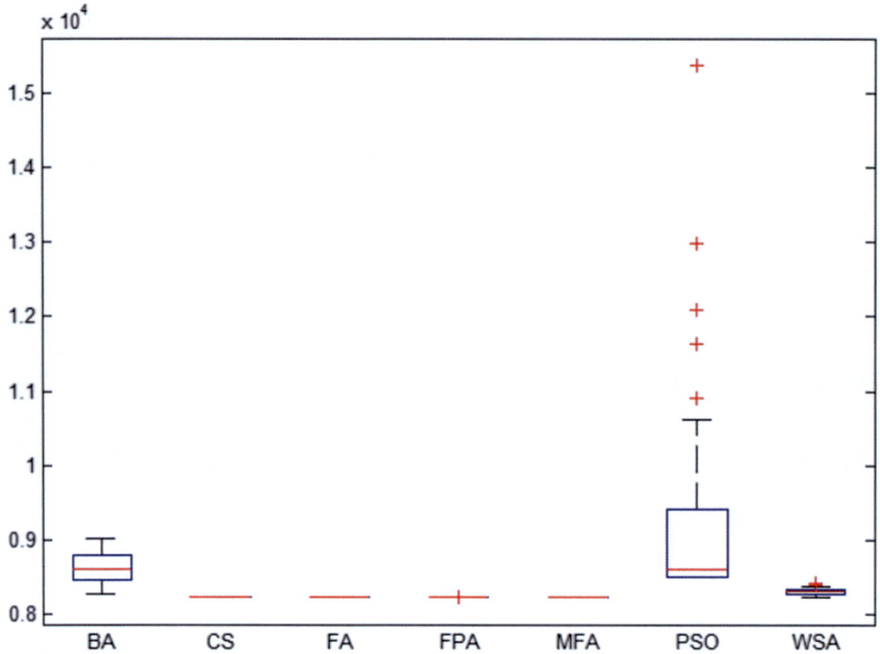

Fig. 2 The fuel cost in case 2 (50 times)

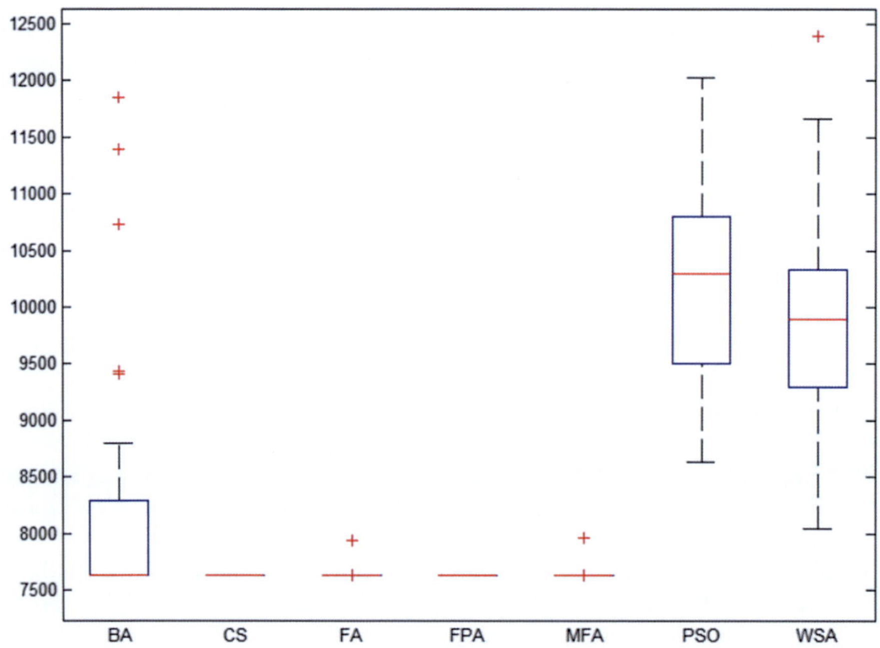

Fig. 3 The generation cost in case 3 (50 times)

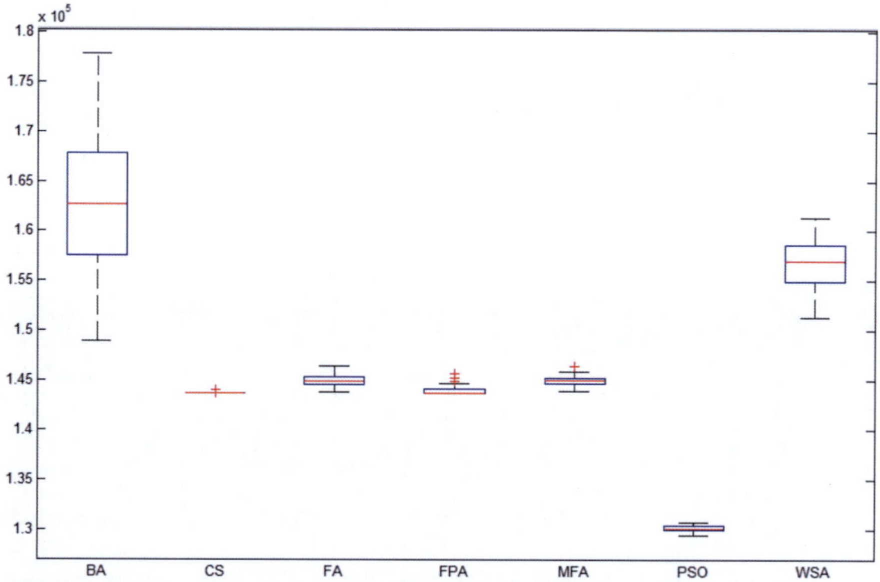

Fig. 4 The generation cost in case 4 (50 times)

The Economic Load Dispatch Problem—Economic load dispatch is the short-term determination of the optimal output of a number of electricity generation facilities, to meet the system load, at the lowest possible cost, subject to transmission and operational constraints. The Economic Dispatch Problem is solved by specialized computer software that should satisfy the operational and system constraints of the available resources and corresponding transmission capabilities. In the US Energy Policy Act of 2005, the term is defined as "the operation of generation facilities to produce energy at the lowest cost to reliably serve consumers, recognizing any operational limits of generation and transmission facilities".

The main idea is that, in order to satisfy the load at a minimum total cost, the set of generators with the lowest marginal costs must be used first, with the marginal cost of the final generator needed to meet load setting the system marginal cost. This is the cost of delivering one additional MWh of energy onto the system. The historic methodology for economic dispatch was developed to manage fossil fuel burning power plants, relying on calculations involving the input/output characteristics of power stations.

Algorithm—A self-contained step-by-step set of operations to be performed. Algorithms exist that perform calculation, data processing, and automated reasoning.

Table 15 Power needed to produce for each unit under different algorithms in different cases

Units	BA	CS	FA	FPA	MFA	PSO	WSA
Case 1							
1	34.83153	31.94724	31.95595	31.94429	31.95595	30.85748	31.81876
2	67.44427	67.28644	67.31757	67.28794	67.31757	66.52586	67.22075
3	47.75473	50.79685	50.757	50.7983	50.757	52.64718	50.99102
Case 2							
1	261.9555	312.713	312.8271	312.7789	312.8271	199.9977	284.231
2	50.13258	72.52534	72.42112	72.43884	72.42112	100.0022	67.20761
3	171.6255	159.8879	159.9221	159.9028	159.9221	100	159.1629
4	59.73941	50	50	50.00073	50	100	50.04063
5	102.8283	54.87384	54.8296	54.87865	54.8296	100	89.35782
6	53.7187	50	50.00003	50.00014	50.00003	100	50.00011
Case 3							
1	0	0	0	0	0	0	6.64E−15
2	0	0	0	3.25E−15	0	0	1.30E−14
3	0	0	0	9.38E−15	0	0	1.58E−14
4	60	60	60	60	60	59.96381	60
5	60	60	60	60	60	60.00089	60
6	60	60	60	60	60	59.93879	109.8666
7	60	60	60	60	60	108.4305	60
8	60	60	60	60	60	109.7468	60
9	60	60	60	60	60	159.5849	60
10	40	40	40	40	40	38.94841	40
11	40	40	40	40	40	31.94568	40
12	55	55	55	55	55	13.71798	55
13	55	55	55	55	55	13.76123	55
Case 4							
1	36	36	37.89602	36	36.01008	73.46373	36.13568
2	114	36	36.8527	36	36.05942	36.30088	73.09138
3	120	60	60.24098	60	60.21223	59.8508	60
4	80	80	80.00161	80	80	75.17039	80.00177
5	47	47	47.6023	47	55.23361	46.7947	47.00003
6	68	68	68.106	68	68.02133	50.24505	68
7	110	110	110	110	110	103.7647	110.0154
8	300	135	135	135	135.0654	63.82188	135
9	135	135	135.0167	135	135	65.74083	135.2382
10	130	130	130.0004	130	130	50.93811	130.0251

(continued)

Table 15 (continued)

Units	BA	CS	FA	FPA	MFA	PSO	WSA
11	94	94	94.00266	94	94	61.45152	94.28999
12	94	94	94	94	94	67.62147	94.00023
13	125	125	125	125	125	76.25026	125.0001
14	125	125	125	125	125.0014	112.0129	210.7708
15	125	125	125.0001	125	125	107.2215	125.0024
16	125	125	125	125	125.0003	101.0213	125.1409
17	220	220	220.0032	220	220	99.47135	309.7598
18	220	220	220	220	220.0041	100.5599	220
19	242	242	242.0017	242	242	62.27842	242.0383
20	242	242	242.0104	242	242	62.50266	242.062
21	254	254	254	254	254	74.12053	342.6853
22	254	254	254.0061	254	254.0017	74.38525	254
23	254	254	254.0054	254	254	73.85946	420.7612
24	254	254	254	254	254	74.16868	254.0006
25	254	254	254	254	254	74.66109	343.7594
26	254	254	254.0059	254	254.0059	73.90539	343.1861
27	10	10	10.00065	10	10.00702	36.08791	10.00492
28	10	10	10	10	10.01851	36	10.00005
29	10	10	10.009	10	10.00356	36.18478	10.00118
30	47	47	49.84489	47	48.3053	46.40879	47.0006
31	60	60	60.00631	60	60.02181	58.52271	60.00007
32	60	60	60.00463	60	60.04994	59.65099	109.7315
33	190	60	60.0022	60	60.00436	59.73774	107.3949
34	90	90	90.36668	90	90.28313	85.37732	90.00008
35	90	90	90.0792	90	90.28334	85.52529	90
36	200	90	90.00499	90	90.07707	89.18732	90.11148
37	25	25	25.00147	25	25.14367	36.14299	25.00001
38	25	25	25.28209	25	25.01891	60.7218	25.00004
39	25	25	25.00754	25	25.47669	48.75879	25.50306
40	242	242	242	242	242	62.22651	310.9445

Appendix

See Tables 15 and 16.

Table 16 The algorithms to be compared introduction

Algorithm	Author	Year	Nature behavior	Unique character
PSO	Kenneth and Eberhart	1995	Swarming behavior of fish and birds	Mainly mutation and selection, high degree of exploration; converging quickly
FA	Xin-She Yang	2008	Flashing behavior of swarming fireflies	Attraction is used that is seeking the optimum by subdivided group; Dealing with the multimodal problems well
MFA	Xin-She Yang	2015	Greedy idea added base on FA	Helping FA find more converged solution
CS	Xin-She Yang, Suash Deb	2009	Brooding parasitism of cuckoo	Bing enhanced by the so-called Levy-flight; efficient random walks and balanced mixing and very efficient in global search
BA	Xin-She Yang	2010	Eholocation of forging bat	Frequency tuning is firstly used; the Mutation can vary due to the variations of the bat loudness and pulse emission
PFA	Xin-She Yang	2012	Flower pollination (mutation) activities	mutation activities can occur at all scales, both local and global; Having extended to multi-objective problems
WSA	Tang Rui, Simon Fong	2012	Wolf hunting and escaping behavior	Having the local mutation and use the jump probability to avoid being caught in the local modal

References

Chen, P.-H., & Chang, H.-C. (1995). Large-scale economic dispatch by genetic algorithm. *IEEE Transactions on Power Systems, 10,* 1919–1926.

Gao, M. L., Li, L. L., Sun, X. M., & Luo, D. S. (2015). Firefly algorithm (FA) based particle filter method for visual tracking. *Optik, 126,* 1705–1711.

http://www.mathworks.com/matlabcentral/fileexchange/7506-particle-swarm-optimization-toolbox.

http://www.mathworks.com/matlabcentral/fileexchange/29693-firefly-algorithm.

http://www.mathworks.com/matlabcentral/fileexchange/37582-bat-algorithm–demo.

http://www.mathworks.com/matlabcentral/fileexchange/45112-flower-pollination-algorithm.
Kennedy, J., & Eberhart, R. (1995). Particle swarm optimization. In *Proceedings of IEEE International Conference on Neural Networks* IV, pp. 1942–1948.
Saadat, H. (1999). *Power System Analysis*. McGraw-Hill companies, Inc.
Sinha, N., Chakrabarti, R., & Chattopadhyay, P. K. (2003). Evolutionary programming techniques for economic load dispatch. *IEEE Transactions on Evolutionary Computation, 7*(1), 83–94.
Tang, R., Fong, S., Yang, X. S., & Deb, S. (2012). Wolf search algorithm with ephemeral memory. In *2012 Seventh International Conference on Digital Information Management (ICDIM)*, pp. 165, 172, August 22–24, 2012.
Yang, X. S. (2009). Firefly algorithms for multimodal optimization. In Watanabe, O., & Zeugmann, T. (Eds.), *Stochastic Algorithms: Foundations and Applications, GA2009, Lecture Notes in Computer Science* (vol. 5792, pp. 169–178). Berlin: Springer.
Yang, X. S. (2010a). A new metaheuristic bat-inspired algorithm. In *Nature Inspired Cooperative Strategies for Optimization (NICSO 2010)*, pp. 65–74.
Yang, X.-S. (2010b). *Nature-Inspired Metaheuristic Algorithms* (2nd ed). Luniver Press.
Yang, X.-S. (2012). Flower pollination algorithm for global optimization. In *International Conference on Unconventional Computing and Natural Computation, UCNC 2012*, pp. 240–249.
Yang, X.-S., & Deb, S. (2009). Cuckoo search via levy flights. In *Nature and Biologically Inspired Computing, 2009. World Congress on NaBIC 2009* (pp. 210–214). IEEE.
Yang, X. S., Hosseini, S. S., & Gandomi, A. H. (2012). Firefly algorithm for solving non-convex economic dispatch problems with valve loading effect. *Applied Soft Computing, 12*(3), 1180–1186. ISSN 1568-4946.
Zaraki, A., & Bin Othman, M. F. (2009). Implementing particle swarm optimization to solve economic load dispatch problem. In *International Conference of Soft Computing and Pattern Recognition, 2009. SOCPAR '09*, pp. 60, 65, December 4–7, 2009. https://doi.org/10.1109/socpar.2009.2.

Dr. Simon Fong graduated from La Trobe University, Australia, with a 1st Class Honours B.E. Computer Systems degree and a Ph.D. Computer Science degree in 1993 and 1998, respectively. Simon is now working as an Associate Professor at the Computer and Information Science Department of the University of Macau. He is a co-founder of the Data Analytics and Collaborative Computing Research Group in the Faculty of Science and Technology. Prior to his academic career, Simon took up various managerial and technical posts, such as systems engineer, IT consultant, and e-commerce director in Australia and Asia. Dr. Fong has published over 432 international conference and peer-reviewed journal papers, mostly in the areas of data mining, data stream mining, big data analytics, meta-heuristics optimization algorithms, and their applications. He serves on the editorial boards of the Journal of Network and Computer Applications of Elsevier (I.F. 3.5), IEEE IT Professional Magazine, (I.F. 1.661), and various special issues of SCIE-indexed journals. Simon is also an active researcher with leading positions such as Vice-chair of IEEE Computational Intelligence Society (CIS) Task Force on "Business Intelligence and Knowledge Management", and Vice-director of International Consortium for Optimization and Modeling in Science and Industry (iCOMSI).

Ms. Tengyue Li is currently an M.Sc student major in E-Commerce Technology at the Department of Computer and Information Science, University of Macau, Macau SAR of China. She participated in the university the following activities: Advanced Individual in the School, Second Prize in the Smart City APP Design Competition of Macau, and Top 10 in the China Banking Cup Million Venture Contest. Campus Ambassador of Love. Tengyue has internship experiences as aa Meituan Technology Company Product Manager from June to August 2017. She worked at the Training Base of Huawei Technologies Co., Ltd. from September to October 2016. From February to June 2016, Tengyue worked at Beijing Yangguang Shengda Network Communications as a data analyst. Lately, Tengyue involved in projects such as "A Minutes" Unmanned Supermarket by the University of Macau Incubation Venture Project since September 2017.

Ms. Zhiyan Qu is a former M.Sc student major in E-Commerce Technology at the Department of Computer and Information Science, University of Macau, Macau SAR of China. Zhiyan completed her study in mid-2018.

Chapter 8
The Paradigm of Fog Computing with Bio-inspired Search Methods and the "5Vs" of Big Data

Richard Millham, Israel Edem Agbehadji, and Samuel Ofori Frimpong

1 Introduction

Big data is a paradigm of large availability of data that is created every second. It is estimated that 1.7 billion messages are created each day from social media big data platforms (Patel et al. 2014). Social media is a platform where people share opinions and thoughts. In view of this, organizations and businesses are overwhelmed by the amount of data and the variety of data cascading through their operations as they struggle to store the data—much less analyze, interpret and present it in meaningful ways (Intel 2013). Thus, most big data yields neither meaning nor value and to find the underlying causes, it is important to understand the unique features of data, namely high dimensionality, heterogeneousness, complexity, unstructuredness, incompleteness, noisiness and erroneousness, which may change the data analysis approach and its underlying statistical techniques (Ma et al. 2014).

The proliferation of Internet of things (IoT) and sensor-based applications have also contributed to having data with different features from different "things" connected together. Consequently, data analytics frameworks have to be re-examined from the 5Vs' (that is velocity, variety, veracity, volume and value) perspective data to determine the "essential characteristics" and the "quality-of-use".

IoT has revolutionized ubiquitous computing and has made several applications to be built around different kinds of sensors. For instance, vast activities are seen

R. Millham (✉) · I. E. Agbehadji · S. O. Frimpong
ICT and Society Research Group, Durban University of Technology, Durban, South Africa
e-mail: richardm1@dut.ac.za

I. E. Agbehadji
e-mail: israeldel2006@gmail.com

S. O. Frimpong
e-mail: Spiritus83@yahoo.com

S. J. Fong and R. C. Millham (eds.), *Bio-inspired Algorithms for Data Streaming and Visualization, Big Data Management, and Fog Computing*, Springer Tracts in Nature-Inspired Computing,
https://doi.org/10.1007/978-981-15-6695-0_8

in IoT-based product-lines of some industries. These activities are expected to grow with projections as high as billions of devices with on average 6–7 devices per person (Ejaz et al. 2016). Consequently, IoT will generate more data, and the data transfer to and from IoT devices will increase substantially. IoT devices refer to the devices that have sensing and actuating capability (Naha et al. 2018). The transfer to and from IoT devices is a challenge to data analytics platforms because it is unable to process huge amount of data quickly and accurately which may affect the "quality-of-use" of data in decision making. Additionally, big data analytics framework creates bottleneck during processing and communication of data (Tsai et al. 2015). This bottleneck needs to be addressed in order to uncover the full potential of IoT. Thus, it is significant to re-design data analytics framework in order to improve on "quality-of-use" of data taking into consideration the "essential characteristics" of an IoT devices such as the 5V's.

Singh and Singh (2012) state that the challenges of big data are data variety, volume and analytical workload complexity. Therefore, organizations that use big data need to reduce the amount of data being stored as it could improve performance and storage utilization. This indicates that variety and volume are essential to improve performance and storage on big data platforms. Additionally, when the essential attributes are addressed, it could reduce the workload complexity of data analytics platforms.

Fog computing paradigm focuses on devices connected to the edge of networks (that is, switches, routers, server nodes). The term fog computing or edge computing operates on the concept that instead of hosting devices to work from a centralized location that is cloud server, fog systems operate on network ends (Naha et al. 2018). Fog computing architecture plays a significant role in big data analysis in terms of managing the large variety, volume and velocity of data from IoT devices and sensors connected to fog computing platform. The platforms manage applications within the fog environment in terms of allocating resource to users, scheduling resources, fault tolerance, "multi-tenancy", security of application and users data (Naha et al. 2018). Basically, fog computing avoids delay in processing of raw data collected from edge networks. Afterward, the processed data is transmitted to cloud computing platform for permanent storage. Additionally, fog computing architecture manages the energy required to process raw data from IoT devices and sensors; thus, optimizing energy requirement is important for data processing in fog computing. Therefore, fog computing monitors the Quality of Service (QoS) and "Quality of Energy" (QoE) in real time and then adjusts the service demands (Pooranian et al. 2017).

The benefit of IoT and big data initiatives are many. For instance, big data and IoT have been used in health sector to monitor quality of service delivery; governments can use it to reach out to its citizenry for better social intervention programs; companies have use it to understand its customers perceptions on products, optimize organizational process and activities to deliver quality service. Similarly, businesses can apply it in cases of remote and on-site members on a project, where each mobile on-site member easily explores data, discovers hidden trends and patterns and communicates their findings to remote sites.

2 '5Vs' of Big Data

The "5Vs" of big data refers to characteristics such as volume, velocity, variety, veracity and value. Although there are several characteristics, in this chapter we consider the predominant "5Vs" as enablers of IoT data.

2.1 *Volume Characteristics*

Volume refers to the amount of data. Handheld devices generate high volume of data as a result of user interactions. When a user inputs data, the handheld Internet-enabled devices send the data for further analysis on the data analytics platform. The challenge that might be created is bottleneck as several devices may compete for processing and communication structure. Because of the competition, handheld Internet-enabled device is an essential component which should be considered by data analytics frameworks. One challenge created when multiple handheld Internet-enabled devices, such as sensors, each send raw data, which they gathered, to the big data analytics framework is that bottlenecks are quickly created in the processing and communication structure. (Tsai et al. 2015). Consequently, there is a need to avoid these bottlenecks by shifting some of the processing down near the level of sensors in the form of fog computing.

2.2 *Velocity Characteristics*

Velocity is the rate of data transfer from Internet-enabled or sensor-enabled handheld devices. As long as devices are connected to Internet, data is processed in real time. However, "real-time or streaming data bring up the problem of large quantity of data coming into the data analytics within a short duration and, the device and system may not be able to handle these input data" (Tsai et al. 2015).

2.3 *Variety Characteristics*

Variety is the different type of data sent by a user. The data could be in the form of text, picture, video and audio. There are devices that are specially designed to handle different forms of data. In most instances, user devices are adapted to handle multiple types of data. This means that processing frameworks should be developed to identify and separate different kinds of data. To achieve this, different classification algorithms and clustering algorithms would help to identify and separate data when the fog computing framework is used.

2.4 Veracity Characteristics

Veracity is the level of quality, accuracy and uncertainty of data and data sources. Veracity is equally associated with how trustworthy the data is. Mostly, location of IoT devices and landmark information can increase trustworthiness. The application of fog computing framework could help to process the data by determining the exact location of data sources. The exact location can be determined by applying location based algorithms (Lei et al. 2017).

2.5 Value Characteristics

Value appears at the final stage of the proposed model. The value of data can refer to the important feature of data that gives value to a business process and activity (Hadi et al. 2015). The business value could be in terms of different opportunities of revenue, creating an innovative market, improving customer experiences etc. (Intel 2013). Singh and Singh (2012) indicates that using big data in healthcare plans of US can drive efficiency and quality, and create an estimated amount of US$300 billion in value each year by reducing healthcare expenditure by 8%. Similarly, developed European economies could save approximately US$149 billion in value on operational efficiency improvement.

In view of these "5Vs" discussed, fog computing plays a key role in managing these characteristics which is discussed in subsequent paragraphs.

3 Fog Computing

Fog computing is based on the concept of allowing data to be processed at the location of devices instead of directly to a cloud environment which creates a data bottleneck. In general, edge computing does not associate with any types of cloud-based services (Naha et al. 2018). When devices are able to process data directly on fog computing platform it improve performance, guarantee fast response time and avoids delay or jitter (Kum et al. 2017). During the processing of data, the device interacts with an intermediate computing framework referred to as fog computing framework/architecture. The fog computing consists of fog server nodes which devices (e.g., IoT devices) are connected to. Fog server is capable of processing data to avoid delay or jitter (Kum et al. 2017). Fog computing is extends the capability of cloud computing. Whereas cloud computing provides "distributed computing and storage capacity", fog computing is closer to the IoT devices/edge of network and help with real-time data analysis (Ma et al. 2018). Thus, fog computing liaise between IoT devices and cloud computing framework. Table 1 shows the attributes that distinguishes cloud computing and fog computing.

Table 1 Attributes to distinguish cloud and fog computing

Cloud attributes	Fog attributes
Vertical resource scaling	Vertical and horizontal resource scaling
Large-size and centralized	Small-size and spatially distributed
Multi-hop wide area network-based access	Single-hop wireless local area network-based access
"High communication latency and service deployment"	Low communication latency and service deployment
"Ubiquitous coverage and fault-resilient"	"Intermittent coverage and fault-sensitive"
"Context-unawareness"	"Context awareness"
Limited support to device mobility	Full support to device mobility
Support to computing-intensive delay-tolerant analytics	Support to real-time streaming applications
"Unlimited power supply" (exploitation of electrical grids)	"Limited power supply" (exploitation of "renewable energy")
Limited support to the device heterogeneity	Full support to the device heterogeneity
Virtual machine-based resource virtualization	Container-based resource virtualization
High inter-application isolation	Reduced inter-application isolation

Source (Baccarelli et al. 2017)

The following are attributes/characteristics of fog computing framework, namely low latency, location identification (fast re-activation of nodes); wide "geographical distribution"; "large number of nodes and mobility", supports IPv6; supports "wireless access"; allows "streaming and real-time" application; and supports "node heterogeneity" (Luntovskyy and Nedashkivskiy 2017). These characteristics provide an ideal platform for the design and deployment of IoT-based services". The advantage of fog computing is the efficient services delivery and reduction in electric energy consumption to transmit data to cloud systems for processing. The fog computing architecture is shown in Fig. 1.

Figure 1 illustrates fog computing architecture as a three-tier network structure. The first tier shows the initial location where Internet-enabled devices are connected. The second tier shows the interconnection of fog devices/nodes (including servers and devices, such as routers, gateways, switches and access points) that is responsible for processing, computing and storing the sensed data temporarily (Yuan et al. 2017). Fog servers that manage several "Fog devices and Fog gateways can translate services between heterogeneous devices (Naha et al. 2018). The upper tier is

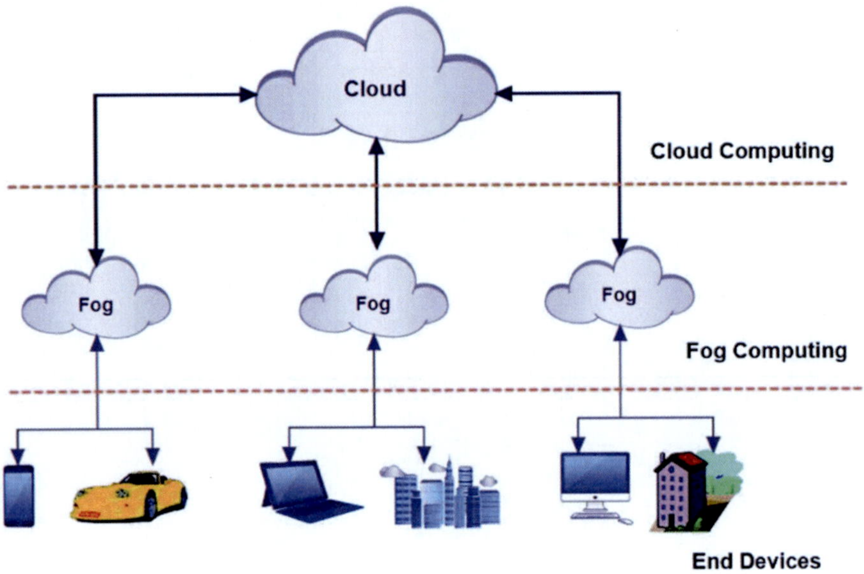

Fig. 1 Fog computing model

the cloud computing layer, which consist of data center-based cloud, "processes and stores an enormous amount of data" (Yuan et al. 2017).

In Fig. 1, it is observed that fog nodes are geographically dispersed in terms of cities, individual houses and dynamically moving vehicles. In view of this, search algorithms are significant to determine the location, scale and resource constraints of each node. One of these algorithms is the Fast Search and "Find of Density" Peak clustering algorithm for load balancing on fog node (Yuan et al. 2017).

The data analytics framework may consist of four layers, namely IOT device, aggregation, processing and analytics layers (Ma et al. 2018). The IoT device layer generates the raw data which is then aggregated (grouped together to reduce dimension/amount of data) before it is sent up to the upper layers for processing and data analysis. The fog computing, which is located on the upper layer, does the processing and transmit the final data to the cloud computing architecture for future storage.

3.1 Fog Computing Models

The following subsection discusses some application of fog computing:

3.1.1 Fog Computing in Smart City Monitoring

Fog computing is applied in "smart cities" to identify anomalous and hazardous events and to offer optimal responses and control. A smart city refers to an urbanized area, where multiple locations of the city "cooperate to achieve sustainable outcomes through the analysis of contextual and real-time information" on events (Tang et al. 2017). In the smart city monitoring model, a hierarchical distributed fog computing framework is used in data analysis to identify and find optimal responses to events (Tang et al. 2017). This hierarchical distributed framework uses fiber optic sensors together with sequential learning algorithm to find anomalous events on pipelines and extract relevant features on events. Basically, pipelines allows resource and energy distribution in smart cities. Therefore, any challenge with pipeline structure threatens the smart cities concepts. The challenges of pipeline include aging and environmental changes which leads to corrosion, leakage and failures of pipeline. In view of this challenge with pipeline, Tang et al. (2017) proposed a four layer fog computing framework to monitor in real-time pipelines and detect three levels of emergency: "Long-term and large-scale emergency events (earthquake, extremely cold or hot weather, etc.)"—referring to level 1 of Data center on the Cloud; "Significant perturbation (damaged pipeline, approaching fire", etc.—referring to level 2 of Intermediate computing nodes; and Disturbances (leakage, corrosion, etc.)—referring to level 1 of the Edge devices. At layer 4, "optical fibers are used as sensors to measure the temperature along the pipeline. Optical frequency domain reflectometry (OFDR) system is applied to measure the discontinuity of the regular optical fibers" (Tang et al. 2017).

A four-layer fog computing architecture in smart cities is possible where the coverage and latency-sensitive applications operate near the edge of the network nearest the sensor infrastructure. This architecture provides very quick response times but does require multiple components as scalability of these components may not be possible. Within this framework, the layer 4 is at the edge of network that is the "sensing network that contains numerous sensory nodes" that are widely "distributed at various public infrastructures to monitor their condition changes over time" (Tang et al. 2017). The data stream from layer 4 is transfer to layer 3 that consists of high performance and low-powered computing edge devices "where each edge device is able to be connected to group of sensors that often encompass a neighborhood and performs data analysis in real-time. The output from an edge device consists of two aspects: the first results of data processing is sent to intermediate computing node at the upper layer. The second is feedback control to a local infrastructure that respond to any threat that may occur on any infrastructure components. The layer 2 consists of several intermediate nodes, each of which is connected to a group of edge devices at layer 3 and associates spatial and temporal data to identify potential hazardous events. Meanwhile, it makes quick response to control the infrastructure when hazardous events are detected. The feedback control that is provided at layers 2 and 3 acts as localized "reflex" decisions to avoid potential damage. For example, if one segment of gas pipeline is experiencing a leakage or a fire is detected, these computing nodes will detect the threat and quickly shutdown the gas supply. Meanwhile, all the data analysis results are sent to the top layer which performs a more complex analysis. The top layer is a cloud computing data center that provides monitoring and centralized control of events. The distributed computing and storage capacity allows large-scale event detection, long-term pattern recognition and relationship modeling that support dynamic decision making (Tang et al. 2017).

The fog computing architecture has significant advantages over the cloud computing architecture in smart city monitoring. This is because, firstly, the distributed computing and storage nodes of fog computing support the massive numbers of sensors distributed throughout a city to monitor infrastructure and environmental parameters. The challenge of only using cloud computing for this smart city monitoring task, is that huge amounts of data will be transmitted to data centers, which results in massive communication bandwidth and power consumption (Tang et al. 2012).

The advantage of smart city is that it can reduce traffic congestion and energy waste, while allocating constraint resources more efficiently to improve quality of life in cities. Smart city also presents many challenges such as how to create an accurate sensing network to monitor the infrastructure components including roads, water pipelines, etc. as urbanization increases; in smart cities large amount of data is generated from sensor networks which leads to big data analytics challenges (Tang et al. 2017); inter-communication between sensor network creates network traffic; and integration infrastructure components and service delivery to ensure efficient control and feedback for decision making. In view of these challenges, fog computing presents a unique opportunity to address these challenges.

In a smart city, there are a number of sensors within sub-units of this city such as IoT homes, energy grids, vehicles and industries. Each unit or group of units would have a proximate aggregating center nearby that groups this data and then communicates and interacts with a computing model to provide an "intelligent" management system for these units. This system then engages with interacting units in order to provide coordination and optimize resources. A remote aggregation center may also be present to group data flowing from sensors in the smart city for further analysis but not for immediate reaction.

3.1.2 Smart Data Model

Hosseinpour et al. (2016) proposed a model to reduce data size generated from IoT sensors via adaptive self-managed and lightweight data cells referred as smart data.

The data, besides the raw data including that sent by the sensors, includes logs and timestamps. The metadata includes information such as the source of data (e.g., sensors), where data is being sent to, the physical entity that data belongs to, timestamps etc. The virtual machine then executes the rules on the metadata using code modules within application software for a particular service. These modules may be an application-specific module, compression module, filtering module, security module, et al. Thus when a service is no longer needed the code becomes in-active in the module structure. This code module are built into smart data cell as plugins which can be enabled using "remote code repository node" that contains all code modules. When a specific code is required by the smart data cell it then sends information to the code repository and the request is granted (Hosseinpour et al. 2016).

In order to avoid communication overhead, recent downloads are cached in the physical fog nodes. Whenever, requested "code module does not exist in the local fog node, it is downloaded from the remote code repository node". The smart data model takes into consideration of the hierarchical structure of fog computing system as it is the main enabler for "implementing smart data concept". The smart data model is controlled and managed through set of rules that defines the behavior of metadata (Hosseinpour et al. 2016).

The advantage of smart data model is that it reduces the "computing load and communication overhead" imposed by the big data. Additionally, it avoids placement of application codes on each fog node then reducing the time of executing which leads to reduce communication overhead cost and energy required for data to move within the fog network (Hosseinpour et al. 2016).

3.1.3 Energy Efficient Model

Oma et al. (2018) proposes an energy efficient model when large number of device, namely sensors and actuators are connected with cloud server in IoT. In this model, when sensors create large image data it is transmitted to cloud servers on the network. The cloud server then decides necessary actions to process the image data and then transmit the actions to an actuator in real time. However, transmitting image data over a network consumes a significant amount of energy. Thus, an intermediate layer (fog layer) is introduced between clouds and devices in IoT. In view of this, the processing and storage of data on server are distributed to fog nodes while permanent data to be stored is transmitted to cloud server.

Oma et al. (2018) energy model is a linear IoT model that "deploy processes and data to devices, fog nodes and servers in IoT" so that total "electric energy consumption of nodes" can be minimized. Although other energy consumption models exist (Enokido et al. 2010, 2011, 2014), Oma et al. (2018) used "simple power consumption (SPC) model" where power consumption of fog nodes are based on maximum and minimum electric energy consumption to process a data size (e.g., data of size x).

An experiment was conducted using a Raspberry Pi as a fog node. This node has "minimum electric power of 2.1 W and the maximum electric power of 3.7 W. The computation rate of each fog node is computed to be approximately 0.185". The computation rate of the server is 0.185. The finding of the study indicates that if a process is performed without any other process on a fog node it takes 4.75 s. The execution time of a process on a fog node is 4.75/ms (Oma et al. 2018).

Similarly, an experiment was conducted in the cloud computing model. During the experiment a sequence of routers from sensor nodes to a server were considered. Each router just forwards messages to another router. Hence, each router supports the input and output. The data obtained from sensors is forwarded from a router to another router (Oma et al. 2018).

Figures 2 and 3 show the "total electric energy expended by nodes" in an IoT model and the cloud model.

Figure 2 shows the sum of execution time of nodes in IoT model.

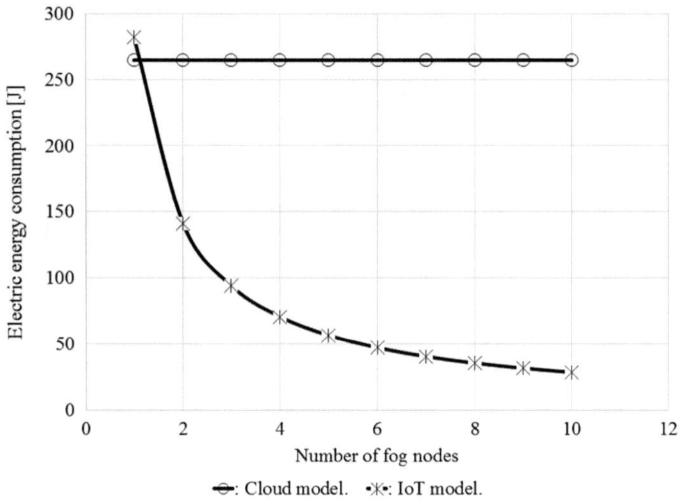

Fig. 2 Electric energy consumption. *Source* (Oma et al. 2018)

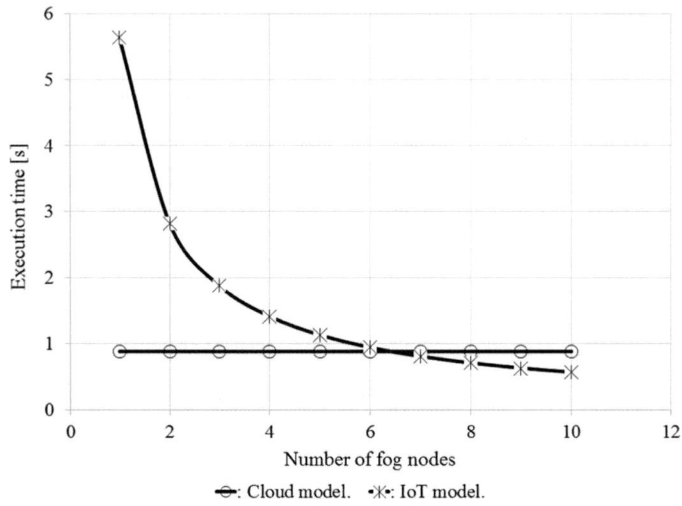

Fig. 3 Execution time. *Source* (Oma et al. 2018)

The experimental results shows that the total electric energy consumed by fog node and a server node, and the total execution time of the linear IoT model is smaller than the cloud computing model (Oma et al. 2018).

3.1.4 Fog Computing for Health Monitoring

Fog computing enables scalable devices for health monitoring. Isa et al. (2018) researched into health related monitoring application where heart of patients are monitored and recorded. Within this monitoring system the electrocardiogram (ECG) signals are processed and analyzed by fog processing units within time constraint. Fog processing unit plays a significant role in "detecting any abnormality in the ECG signal detected. The location of these fog processing servers are important to optimize the energy consumption of both processing and networking equipment" (Isa et al. 2018). The mixed integer linear programming (MILP) approach was adopted to "optimize total energy consumption of the health monitoring system" (Isa et al. 2018).

A GPON architecture with fog network shows device connection from the edge to a central cloud. On this network, there are three layers, namely the "access network, metro network and core network". In the context of health systems, Fog computing can be deployed in two layers (Isa et al. 2018). The first layer is for processing servers (PS) to connect the "Optical Network Terminals (ONT) of the Gigabit Passive Optical Network (GPON)". Therefore, when the processing servers is place on this layer, that is closer to the users, it can reduce energy consumption of a networking equipment. However, it increases the required number of processing servers. On the other hand, the second layer has processing servers connected to the "Optical Line Terminal (OLT)". Therefore, using processing servers reduces the number of required processing servers which is a shared "point between the access points". However, this increases energy consumption of the networked equipment (Isa et al. 2018).

An experiment was conducted to evaluate the model and data was collected from 200 patients uniformly distributed among 32 Wi-fi access points (Isa et al. 2018). The results show that processing the ECG signal saves up to 68% of total energy consumption (Isa et al. 2018).

3.2 *Fog Computing and Swarm Optimization Models*

This section presents on models to address challenges in fog computing when more edge devices are connected. The challenges includes energy consumption, data distribution, heterogeneity of edge devices, dynamicity of fog network etc. as more devices are connected. This leads to finding new methods to address the challenges that were identified.

One of the methods is the use of bio-inspired algorithm. In this regard, researchers have developed different models and methods that combine fog computing and bio-inspired methods to build dynamic optimization models to these challenges. The following presents the paradigm on fog computing combine bio-inspired methods.

3.2.1 Evolutionary Algorithm for Energy Efficient Model

Mebrek et al. (2017) assessed the suitability of fog computing to increase demand of IOT devices. The assessment focused on the energy consumption demand and quality of service to determine the performance of fog computing. The approach formulated the problem of power consumption and delay in fog as an optimization problem which was solved using evolutionary algorithm (EA) to determine energy efficiency. The problem formulation and the proposed solution IGA (Improved Genetic Algorithm). The model was evaluated using three service scenario, namely (a) for instance with static content; (b) fog computing with dynamic content such as video surveillance; and (c) data is not created in fog instance but pre-downloaded to fog instances from the cloud. These scenarios were used to investigate the behavior of the model in terms of energy consumption of IGA. IGA is used to create a preference list for pairing IoT object-fog instances.

In respect of energy consumption, the performance of the IGA algorithm as compared to traditional cloud solution are similar for static content scenario). Thus, the IoT devices do not take full advantage of the fog resources. As the number of objects increases, the fog utilization rises up. Thus, fog computing architecture is able to improve energy consumption very efficiently (Mebrek et al. 2017).

3.2.2 Bio-Inspired Algorithm for Scheduling of Service Requests to Virtual Machine (VMs)

The role of virtual machine in executing rules cannot be overlooked as the efficient method of executing rules can reduce the energy consumption of edge devices. The energy consumption in fog servers relies on user service request to virtual machines (VMs) (Mishra et al. 2018). Service request allocation is a "nondeterministic polynomial time hard problem". Mishra et al. (2018) present a meta-heuristic algorithm for scheduling of service requests to VMs. The algorithm seeks to minimize the energy consumption at fog server while maintaining the quality of services. This algorithm combines particle swarm optimization (PSO), binary PSO and bat algorithm to handle the heterogeneity of service request in the fog computing platform.

The findings suggest that meta-heuristic techniques help to achieve service allocation energy efficiency as well as achieving the desired Quality of service. Since allocation problem in the fog server system does not have "polynomial time algorithms and nature-inspired algorithms that can supply solutions within a reasonable time period". This findings demonstrated that the BAT-based service allocation algorithm outperforms the other PSO and binary PSO algorithms (Mishra et al. 2018).

3.2.3 Bio-Inspired Algorithms and Fog Computing for Intelligent Computing in Logistic Data Center

The era of using IoT, robots and drones has been integrated into operations of factories in the logistic handling. This current trend has revolutionaries operations of factors such that less or no manpower is involved. This revolution is dubbed as era of Industry 4.0 and it leads to applying intelligence computing systems in logistic operations.

Generally, the cloud computing framework provides a structure where data is located and managed in centralized manner. This structure is a logistic data center because of how it allows integration of technologies of IoT and mobile devices for an intelligent logistic data center management. However if there are several technologies it leads to latency in the response time. Lin and Yang (2018) propose a framework to optimize facilities in factory layout of a logistics data center. The objective of this framework is to deploy intelligent computing systems into operation of logistics center. The facilities have connected devices such as edge devices, gateways, fog devices and sensors for real-time computing.

The integer programming models have been applied to reduce installation cost subject to constraints such as "maximal demand capacity, maximal latency time, coverage and maximal capacity of devices". This approach has been applied in solving NP-hard facility location problem. Meta-heuristic search methods have also been applied to enhance computational efficiency of the search algorithms in order ensure good quality solutions. Some approaches includes the use of discrete monkey algorithm (DMA) to search for good quality solutions, genetic algorithm (GA) to increase computational efficiency etc. The discrete monkey algorithm is based on the characteristics of monkey, namely climbing process, watch-jump process, cooperation with other monkey, and crossover-mutation of each monkey, and somersault process of each monkey. When the hybrid DMA-GA model was simulation, it shows high performance in deployment of intelligent computing systems in logistics centers. The performance of each connected device is evaluated using the following cost function,

$$c_f \sum_{(i,j)\in\{s\}\times\Omega_G\cup\Omega_G\times\Omega_F\cup\Omega_F\times\Omega_E} x_{ij}\cdot d_{ij} + c_G\sum_{m\in\Omega_G} g_m + c_F\sum_{n\in\Omega_F} f_n$$
$$+ c_E\sum_{t\in\Omega_E} q_t + K\cdot(\eta_{\text{link}} + \eta_{\text{demand}} + \eta_{\text{latency}} + \eta_{\text{cover}} + \eta_{\text{capacity}})$$

The equation represents a cost function. the various terms in the equation are expressed as follows: the first term is the "cost of the fiber links between the cloud center and gateways, between gateways and fog devices and between fog devices and edge devices; the other three terms are costs of installing the gateways, fog devices and edge devices, respectively", where C_f is cost of the fiber, C_G is cost of installing gateway, C_F is cost of installing fog device, C_E is cost of installing edge; s is index of cloud center, set of potential sites for gateways Ω_G, set of potential sites for fog Ω_F, set of potential sites for edge Ω_E, x_{ij} is a "binary variable to determine if a link exists or not between i and j nodes, f_n is a binary variable to determine if a potential sites for fog device n is selected to a place fog device, q_t is a binary variable to determine if a potential sites for fog device t is selected to a place an edge device; κ is the penalty cost, which is a very large number; η_{link}, η_{demand}, η_{latency}, η_{cover} and η_{capacity} are numbers of violating linkage between the two" layers of fog architecture (Lin and Yang 2018).

A framework of a computing system deployed in a logistics center may consist of a cloud computing center, fog devices, gateways, edge devices and sensing devices in a top-down fashion (Lin and Yang 2018).

An experiment conducted indicates that although the model produced efficient performance results it could not consider factors of the deployment should e.g., data traffic, latency, energy consumption, load balance, heterogeneity, fairness and quality of service (QoS). An aspect of this framework that can be explored is the application machine learning techniques and the hybrid method of meta-heuristics (DMGA).

3.2.4 Ensemble of Swarm Algorithm for Fire-and-Rescue Operations

Ma et al. (2018) propose a model for data streaming mining to monitor gas data generated from chemical sensors. Gas sensors are mostly located at the edge of network and detecting any anomaly to raise an alert in time is significant. As this leads to necessary emergency rescue services. The challenge with gas monitoring is that when alert is not detected early it may lead to death particularly when the gas is hazardous. Thus, integrating with data mining to "analyze the regularity from the gas sensor monitoring measurement will contribute to occupation safety". The proposed recognize "abnormal gas by accumulating and evaluating various types of gas data and CO_2 from urban automatic fire detection system". Although the proposed model is referred as Internet of Breath, it is based on the fog computing framework.

The model related to the Fog analytics is installed at gas sensor gateway where hardware devices can collect data on gas quality continuously. The edge analysis is achieved using a decision tree model that is built from crunching over the continuous data stream. Features selection is tested using C4.5 and "data mining decision tree algorithm (HT)".

An experiment was conducted using 13,910 dataset collected from 16 chemical sensors to test the model. The classification task grouped data into one of six gases at different concentration levels such Ammonia, Ethylene, Acetaldehyde, Ethanol, Acetone and Toluene. Benchmark data stream mining and machine learning (as well as data mining) platform, namely WEKA and MOA–Massive Online Analysis were used for analysis. The two steps in the experiment are: Firstly, the traditional decision tree algorithm (C4.5) is compared with the data mining decision tree algorithm which is referred to as Hoeffding Tree (HT) and with feature selection (FS) search methods on the two classifiers. These FS algorithms includes GA, Bat, firefly, wolf, cuckoo, flower, Bee, Harmony etc. The performance was evaluated using accuracy, TP rate, Kappa, precision, FP rate, recall and F-measure (Ma et al. 2018).

The results of the experiment indicates that C4.5 has high accuracy if the whole data are trained. In Fog computing environment, data are streaming in large amount continuously into a data stream mining model. In this regard, the model must be able to handle incremental learning where model learns from a portion of data at a time. And it updates itself each time new data is uploaded in real-time; the results suggested that FS has greater impact on C4.5. However, FS ameliorate the performance of HT. Fog computing that is based on HT and FS-Harmony search method could guarantee good accuracy, low latency and reasonable data scalability (Ma et al. 2018).

3.2.5 Evolutionary Computation and Epidemic Models for Data Availability in Fog Computing

Fog computing supports the heterogeneity of devices and ensures dynamicity of a network. Dynamicity is when the node on a network assume different roles with the aim of maintaining data on the network. However, this can create a challenge with data availability and dissemination over the network, which can be resolved by evolutionary computation and epidemic models (Andersson and Britton 2012). Vasconcelos et al. (2018) approach to address the "Data Persistence Problem" in the Fog and Mist computing environment (DPPF) is by using two relatively independent sub-problems and models the FMC environment using graphs.

Devices that are used in the evolutionary computation and epidemic model can assume three roles within the infrastructure that are local leaders (LL), local leaders neighbors (LLN) and far away from local leaders (FLL). The LL nodes helps to control the output rate of nodes that has a copy of data in its neighborhood and manage the data replication process based on measured output rate. The LLN nodes direct copy of the data it has received from its LL node. The FLL nodes starts data replication process by making its own copy of data. The choice of which neighbor to replicate is made based on the roulette wheel selection method. In this way, the data must be replicated to a near LL or a region that has low data availability (Vasconcelos et al. 2018).

In order to ensure that the "data diffusion to the control nodes (LL) are spatially distributed within the topology: and avoid concentration of data at single region of the graph, the "epidemiological data model based on the Reed-Frost model" was adopted. The idea is based on the probability of infection depends mainly on two factors: the first factor is the stability function of the node to be contaminated, since, considering that the most stable nodes possessed the probability of this remaining in the network; and the second factor is the spatial distribution of the data to several other location. In order to know the direction of probability, the idea of genetic algorithm was adopted because of the use of roulette wheel for selection of operator (Vasconcelos et al. 2018).

3.2.6 Bio-Inspired Optimization for Job Scheduling in Fog Computing

Bee Life algorithm (BLA) is a bio-inspired algorithm that is based on the behavior of bees in real life environment. Generally, the behavior of Bee is that it waits on the dance area in order to make decision to choose its food source (Karaboga 2005). Bees are adapted to self-organize itself to enrich their food source and also discard poor sources. This behavior is applied to job scheduling for optimal performance, and cost effective service requests by mobile users (Bitam et al. 2018). The proposed BLA aims to find an optimal distribution of set of task among all the fog computing nodes so as to find "trade-off between CPU execution time and allocated memory". The proposed approach is expressed as job scheduling problem in the fog computing environment. The total CPU execution time of all tasks ('r' tasks) assigned to FN_j is:

$$\mathrm{CPU\_Execution\_Time}(\mathrm{FN}_j\mathrm{Tasks}) = \underset{i \in \text{jobs of selected tasks}}{\overset{\text{sum}}{1 \le k \le r}} \left(J\mathrm{Task}_{ik}^{j}.\mathrm{StartTime} + J\mathrm{Task}_{ik}^{j}.\mathrm{ExeTime}\right)$$

where FN_j Tasks represents the task at fog node each fog node. The time for CPU execution of all tasks ('r' tasks), where $J\mathrm{Task}_{ik}^{j}.\mathrm{StartTime}$ represents the starting time of task "k" of a job "i" executed on FN_j and $J\mathrm{Task}_{ik}^{j}.\mathrm{ExeTime}$ is the CPU execution time of this task "k" at FN_j.

The allocated memory to task "k" assigned to FN_j is calculated, as follows:

$$\mathrm{Memory}(\mathrm{FN}_j\mathrm{Tasks}) = \underset{i \in \text{jobs of selected tasks}}{\overset{\max}{1 \le k \le r}} \left(J\mathrm{Task}_{ik}^{j}.\mathrm{AllocatedMemory}\right)$$

where $J\mathrm{Task}_{ik}^{j}$. AllocatedMemory represents allocated memory at task "k" for job "i". FN_jTasks represents the task at fog node for each fog node (Bitam et al. 2018).

Based on the expression on CPU execution time and allocated memory, the job scheduling problem in the fog computing can be expressed as:

$$\mathrm{FNTasks} = \left(\mathrm{FN}_j\mathrm{Tasks}, \ldots, \mathrm{FN}_n\mathrm{Tasks}\right)$$

where FN_jTasks is refers to tasks assigned to the fog node FN. For its assigned jobs, FN_j ensures the execution of its tasks as follows:

$$\mathrm{FN}_j\mathrm{Tasks} = \left(J\mathrm{Task}_{ax}^{j}, J\mathrm{Task}_{by}^{j}, \ldots, J\mathrm{Task}_{ik}^{j}, \ldots, J\mathrm{Task}_{nr}^{j}\right)$$

The execution of tasks by fog node can be viewed as multiple sequential tasks be executed in different layers within a fog node.

The cost function that can be used to evaluate quality of the expected solution (that is, FNTasks) is expressed as a "minimization function" which is used to "measure the optimality of the two objectives, namely CPU execution time and allocated memory size" (Bitam et al. 2018). This cost function is expressed by:

$$\text{Cost\_function(FNTasks)} = \text{Min}\left[\sum_{m}^{j=1}\left(\text{cost\_function}(J\text{Task}_{ik}^{j}, \text{FN}_j)\right)\right]$$

where,

$$\text{Cost\_function}(J\text{Task}_{ik}^{j}, \text{FN}_j) = w_1.\text{CPU\_Execution\_Time}(\text{FN}_j\text{Tasks}) + w_2.\text{Memory}(\text{FN}_j\text{Tasks})$$

where, w_1 and w_2 represents the weighting factors on importance of each of the two evaluated objectives (i.e., CPU execution time and allocated memory).

A flowchart that illustrates the operational flow of the Bees Life algorithm is explained as follows:

Generation of Initial Population The initialization generates "N" individuals that are selected randomly to form the initial population. To evaluate each individual using the cost function.

Stopping Criterion This is mostly pre-determined by a job scheduler.

Optimization Operators of BLA To ensure diversity of individuals in a given population two genetic operators were applied, namely crossover and mutation. ***Crossover*** operation is "applied on two colony individuals called parents which are the queen and a drone". These parents are then combined to form two new individuals called off-springs. ***Mutation*** is a "unary operator which introduces changes into the characteristics of the off-springs resulting from the crossover operation". Therefore, the new offspring will not be different from the original one (Bitam et al. 2018).

Greedy Local Search Approach In the foraging aspect of BLA, greedy approach was applied for local search to ensure optimal solution among the different individuals in the neighborhood of the original individual. In this approach, individual task can be randomly selected to be substituted by another task from the nearest fog node (Bitam et al. 2018).

Performance Evaluation To evaluate the performance of the BLA framework the following two performance evaluation metrics was used (Bitam et al. 2018).

- CPU execution time (measured in second): is defined as the "time between the start and the completion of a given task executed". The time taken "before and after to separate" and combine task is constant since it do not affect the job scheduling process on a node. The CPU execution time can be calculated as follows:

CPU execution time = number of instructions of a task (i.e., clock cycles for a task)/clock rate

- Allocated memory size (measured in byte) is expressed as the total amount of memory (i.e., the main storage unit) of a fog node, devoted to the execution of a given task.

This model was tested and the results shows that BLA outperforms the particle swarm optimization and genetic algorithm in respect of CPU execution time and allocated memory (Bitam et al. 2018).

3.2.7 Prospects of Fog Computing for Smart Sign Language

Fog computing can also be applied to other disciplines such as sign language studies in order to detect patterns of sign, variation and similarity of sign. Sign language is basically used by the deaf community for communication. Akach and Morgan (1997) describes Sign language as a fully fledged natural language developed through use by a Deaf people. Although it is a natural language used by many countries, it is not a universal language but could be used as medium of communication among Deaf people who resides in different countries. Research estimates the total number of Deaf people who use the Sign language worldwide to be 70 million (Deaf 2016). It is presumed that countries have variation of signs thus sign language used in one country might be different or similar. For instance, there is American Sign Language, British Sign Language, South African Sign Language etc. and all these sign languages have variations and similarities of sign.

The possibility is that Fog computing could be applied to detect aspects of patterns of sign, variation and similarities of signs from different countries and help create a smart sign language system. The benefit of a smart sign language system is that it would facilitate communication among deaf people and hearing people.

4 Proposed Framework on Fog Computing and "5Vs" for Quality-of-Use (QoU)

Based on the models that were reviewed, quality of service is an aspect that has not been fully explored. We propose an analytical model that consider speed, size and type of data from IoT devices and then determine the quality and importance of data to store on cloud platform. This reduces the size of data to store on the cloud platform. The framework is shown in Fig. 4.

Figure 4 shows the design of the proposed framework. The framework has two components, namely IoT (data) and fog computing. The IoT (data) components is the location of sensors, Internet-enabled devices which capture large data, at a speed and different types of data. The challenges of device connected to IoT (data) component includes the energy consumption which is extensively addressed by Mebrek et al.

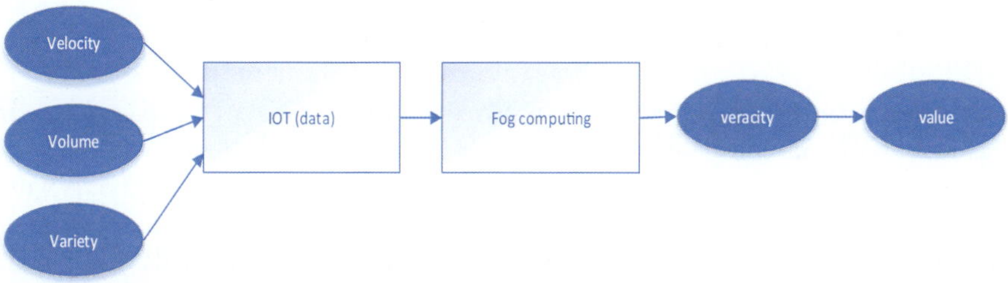

Fig. 4 Proposed framework for IoT (data) for fog computing

(2017) and Oma et al. (2018). The data generated are processed and analyzed by fog computing component to produce quality data that is useful. The quality data and importance (useful) are the attributes of "quality-of-use" which represents the outcome from the framework. This "quality-of-use" characteristics of data shows the added-value that is used for making decision in smart cities reference model.

As earlier shown on Fig. 1, each geographically placed fog node produces different "quality-of-use" dimension. These "quality-of-use" dimensions could be measured through the use of a set of metrics. Additionally, expert knowledge could be applied in selection of the "most valued quality-of-use" data. Although expert knowledge is subjective, it gives a narrow perspective from large volume of data.

In summary, Tables 2 and 3 show attributes of the proposed data analytics framework. The "essential characteristics" are the input attributes, while the "quality-of-use" is the outcome of data.

Although, this model has not been evaluated on real world scenario, it is anticipated that the proposed framework only discover the important data to be stored on the cloud architecture.

Table 2 Essential characteristics

Attributes on essential characteristics	Focus	Description
Volume	Size of data	Quantity of collected and stored data
Velocity	Speed of data	The rate of data transfer between source and destination
Variety	Type of data	The different type of data, namely pictures, videos and audio that arrives at a receiving end

Table 3 Quality-of-use characteristics

Attributes on quality-of-use characteristics	Focus	Description
Value	Importance of data	This represents the business value to be derived from big data, e.g., profit
Veracity	Data quality	Accurate analysis of captured data

5 Conclusion

In this chapter, the following were discussed, namely the "5Vs" of big; fog computing and proposed analytics framework on IoT big data. The challenge with analytics framework is the workload complexity from large volume of data moving with a velocity and with different type of data. The workload creates bottleneck at the processing and communication layers of data analytics platforms. This result in lack of accuracy and latency in sending and capturing of data. The fog computing framework helps to improve accuracy and latency of data. In chapter, we proposed a data analytics framework that combines IoT data and fog computing framework to help reduce workload complexity of data analytics platforms. The approach categorized velocity, volume and variety as "essential characteristics" of IoT devices. Meaning each IoT device captures data with speed, generates large volume of data and with different types of data. Whereas, the fog computing framework analyzes the data to determine the "quality-of-use". The "quality-of-use" data has characteristics of veracity and value. Although, the proposed model is yet to be tested, it is envisaged to reduce the amount of data stored on cloud computing platform. Additionally, this could improve performance and storage utilization problem identified by Singh and Singh (2012).

Key Terminology and Definitions

Fog Computing—Fog computing, also known as fog networking or fogging, is a decentralized computing infrastructure in which data, compute, storage and applications are distributed in the most logical, efficient place between the data source and the cloud. Fog computing essentially extends cloud computing and services to the edge of the network, bringing the advantages and power of the cloud closer to where data is created and acted upon.

Bio-inspired—refers to an approach that mimics the social behavior of birds/animals. Bio-inspired search algorithms may be characterized by randomization, efficient local searches, and the discovering of the global best possible solution.

5Vs' of big data—refers to volume, velocity, variety, veracity and value characteristics of data.

IoT—refers to Internet of things. The "things" refers to Internet-enabled devices that send data over the Internet for processing and analysis. Sensor-enabled devices can also be categorized as "things" that can send data over Internet.

References

Akach, P., & Morgan, R. (1997). *Community interpreting: Sign language interpreting in South Africa.* Paper presented at the Community Interpreting Symposium. Univerity of the Orange Free State, Bloemfontein.

Andersson, H., & Britton, T. (2012). *Stochastic epidemic models and their statistical analysis.* Lecture notes in statistics. New York: Springer.

Baccarelli, E., Naranjo, P. G. V., Scarpiniti, M., Shojafar, M., & Abawajy, J. H. (2017). Fog of everything: Energy-efficient networked computing architectures, research challenges, and a case study.

Bitam, S., Zeadally, S., & Mellouk, A. (2018). Fog computing job scheduling optimization based on bees swarm. *Enterprise Information Systems, 12*(4), 373–397.

Deaf, W. F. O. (2016). *Sign language.* Available https://wfdeaf.org/human-rights/crpd/sign-language/.

Ejaz, W., Anpalagan, A., Imran, M. A., Jo, M., Naeem, M., Qaisar, S. B., et al. (2016). Internet of things (IoT) in 5G wireless communications. *IEEE, 4,* 10310–10314.

Enokido, T., Aikebaier, A., & Takizawa, M. (2010). A model for reducing power consumption in peer-to-peer systems. *IEEE Systems Journal, 4*(2), 221–229.

Enokido, T., Aikebaier, A., & Takizawa, M. (2011). Process allocation algorithms for saving power consumption in peerto-peer systems. *IEEE Transactions on Industrial Electronics, 58*(6), 2097–2105.

Enokido, T., Aikebaier, A., & Takizawa, M. (2014). An extended simple power consumption model for selecting a server to perform computation type processes in digital ecosystems. *IEEE Transactions on Industrial Informatics, 10*(2), 1627–1636.

Hadi, H. J., Shnain, A. H., Hadishaheed, S., & Ahmad, A. H. (2015). Big data and 5v's characteristics. *International Journal of Advances in Electronics and Computer Science, 2*(1), 8.

Hosseinpour, F., Plosila, J., & Tenhunen, H. (2016). An approach for smart management of big data in the fog computing context. In *2016 IEEE 8th International Conference on Cloud Computing Technology and Science* (pp. 468–471).

Intel. (2013). *White Paper, Turning big data into big insights, The rise of visualization-based data discovery tools.*

Isa, I. S. M., Musa, M. O. I., El-Gorashi, T. E. H., Lawey, A. Q., & Elmirghani, J. M. H. (2018). Energy efficiency of fog computing health monitoring applications. In *2018 20th International Conference on Transparent Optical Networks (ICTON)* (pp. 1–5).

Karaboga, D. (2005). An ideal based on honey bee swarm for numerical optimization technical report.

Kum, S. W., Moon, J., & Lim, T.-B. (2017). Design of fog computing based IoT application architecture. In *2017 IEEE 7th International Conference on Consumer Electronics-Berlin (ICCE-Berlin)* (pp. 88-89).

Lei, B., Zhanquan, W., Sun, H., & Huang, S. (2017). Location recommendation algorithm for online social networks based on location trust (p. 6). IEEE.

Lin, C.-C., & Yang, J.-W. (2018). Cost-efficient deployment of fog computing systems at logistics centers in industry 4.0. *IEEE Transactions on Industrial Informatics, 14*(10), 4603–4611.

Luntovskyy, A., & Nedashkivskiy, O. (2017). Intelligent networking and bio-inspired engineering. In *2017 International Conference on Information and Telecommunication Technologies and Radio Electronics (UkrMiCo)*, Odessa, Ukraine (pp. 1–4).

Ma, B. B., Fong, S., & Millham, R. (2018). Data stream mining in fog computing environment with feature selection using ensemble of swarm search algorithms. In *Conference on Information Communications Technology and Society (ICTAS)* (p. 6).

Ma, C., Zhang, H. H., & Wang, X. (2014). Machine learning for big data analytics in plants. *Trends in Plant Science, 19*(12), 798–808.

Mebrek, A., Merghem-Boulahia, L., & Esseghir, M. (2017). Efficient green solution for a balanced energy consumption and delay in the IoT-fog-cloud computing (pp. 1–4). IEEE.

Mishra, S. K., Puthal, D., Rodrigues, J. J. P. C., Sahoo, B., & Dutkiewicz, E. (2018). Sustainable service allocation using a metaheuristic technique in a fog server for industrial applications. *IEEE Transactions on Industrial Informatics, 14*(10), 4497–4506.

Naha, R. K., Garg, S., Georgekopolous, D., Jayaraman, P. P., Gao, L., Xiang, Y., & Ranjan, R. (2018). Fog computing: Survey of trends, architectures, requirements, and research directions. 1–31.

Oma, R., Nakamura, S., Enokido, T., & Takizawa, M. (2018). An energy-efficient model of fog and device nodes in IoT. in *2018 32nd International Conference on Advanced Information Networking and Applications Workshops* (pp. 301–308). IEEE.

Patel, A., Gheewala, H., & Nagla, L. (2014). Using social big media for customer analytics (pp. 1–6). IEEE.

Pooranian, Z., Shojafar, M., Naranjo, P. G. V., Chiaraviglio, L., & Conti, M. (2017). A novel distributed fog-based networked architecture to preserve energy in fog data centers. In *2017 IEEE 14th International Conference on Mobile Ad Hoc and Sensor Systems* (pp. 604–609).

Singh, S., & Singh, N. (2012). Big data analytics. In *International Conference on Communication, Information and Computing Technology (ICCICT)* (pp. 1–4). IEEE.

Tang, B., Chen, Z., Hefferman, G., Pei, S., Wei, T., & He, H. (2017). Incorporating intelligence in fog computing for big data analysis in smart cities. *IEEE Transactions on Industrial Informatics, 13*(5).

Tang, R., Fong, S., Yang, X.-S., & Deb, S. (2012). Integrating nature-inspired optimization algorithms to K-means clustering. In *2012 Seventh International Conference on Digital Information Management (ICDIM)* (pp. 116–123). IEEE.

Tsai, C.-W., Lai, C.-F., Chao, H.-C., & Vasilakos, A. V. (2015). Big data analytics. *Journal of Big data.*

Vasconcelos, D. R., Severino, V. S., Maia, M. E. F., Andrade, R. M. C., & Souza, J. N. (2018). Bio-inspired model for data distribution in fog and mist computing. In *2018 42nd IEEE International Conference on Computer Software & Applications* (pp. 777–782).

Yuan, X., He, Y., Fang, Q., Tong, X., Du, C., & Ding, Y. (2017). An improved fast search and find of density peaks-based fog node location of fog computing system. In *IEEE International Conference on Internet of Things (iThings) and IEEE Green Computing and Communications (GreenCom) and IEEE Cyber, Physical and Social Computing (CPSCom) and IEEE Smart Data (SmartData).*

Richard Millham is currently Associate Professor at Durban University of Technology in Durban, South Africa. After thirteen years of industrial experience, he switched to academe and has worked at universities in Ghana, South Sudan, Scotland and Bahamas. His research interests include software evolution, aspects of cloud computing with m-interaction, big data, data streaming, fog and edge analytics and aspects of the Internet of things. He is a chartered engineer (UK), a chartered engineer assessor and senior member of IEEE.

Israel Edem Agbehadji graduated from the Catholic University College of Ghana with B.Sc. Computer Science in 2007, M.Sc. Industrial Mathematics from the Kwame Nkrumah University of Science and Technology in 2011 and Ph.D. Information Technology from Durban University of Technology (DUT), South Africa, in 2019. He is a member of ICT Society of DUT Research group in the Faculty of Accounting and Informatics; and *IEEE* member. He lectured undergraduate courses in both DUT, South Africa, and a private university, Ghana. Also, he supervised several undergraduate research projects. Prior to his academic career, he took up various managerial positions as the management information systems manager for National Health Insurance Scheme; the postgraduate degree program manager in a private university in Ghana. Currently, he works as Postdoctoral Research Fellow, DUT-South Africa, on joint collaboration research project between South Africa and South Korea. His research interests include big data analytics, Internet of things (IoT), fog computing and optimization algorithms.

Samuel Ofori Frimpong holds a master's degree in Information Technology from Open University Malaysia (2013) and Bachelor of Science degree in Computer Science from Catholic University College of Ghana (2007). Currently, he is a Ph.D. student at the Durban University of Technology, Durban-South Africa. His research interests include Internet of things (IoT) and fog computing.

Chapter 9
Approach to Sentiment Analysis and Business Communication on Social Media

Israel Edem Agbehadji and Abosede Ijabadeniyi

1 Introduction

Social media is an instrument used for communication. This instrument has evolved as medium of social interaction where users can share and re-share information with millions of people. It is estimated that 1.7 billion people use social media to receive or send messages daily (Patel et al. 2014). This shows that many people are using this media to express thoughts and opinions on any subject matter every day. Opinion is transitional concept that reflects attitudes towards an entity (Medhat et al. 2014). The thoughts and opinions are expressed either explicitly or implicitly (Liu 2007). While explicit expression is direct expression of the opinion and thoughts, implicit expression is when a sentence implies an opinion (Kasture and Bhilare 2015). Thus, thoughts and opinions combine explicit and implicit expressions which make its analysis a difficult task. Sentiment analysis is the process of extracting feeling, attitudes or emotion of people from communication (either verbal or non-verbal) (Kasture and Bhilare 2015). The sentiment relates to feeling or emotion, whereas emotion relates to attitude (Mikalai and Themis 2012). Theoretically, sentiment analysis is a field of natural language processing that helps to understand the emotion of humans as they interact with each other via text (Stojanovski et al. 2015). "Natural language processing" (NLP) is a "field of computer science and artificial intelligence that deals with human–computer language interaction" (Devika et al. 2016). Usually, people express their feeling and attitudes using text/words during communication. Thus,

I. E. Agbehadji (✉)
ICT and Society Research Group, Durban University of Technology, Durban, South Africa
e-mail: israeldel2006@gmail.com

A. Ijabadeniyi
International Association for Impact Assessment (Member), Environmental Learning Research Centre, Rhodes University, Grahamstown, South Africa
e-mail: bosede55@yahoo.com

S. J. Fong and R. C. Millham (eds.), *Bio-inspired Algorithms for Data Streaming and Visualization, Big Data Management, and Fog Computing*, Springer Tracts in Nature-Inspired Computing,
https://doi.org/10.1007/978-981-15-6695-0_9

sentiment has three aspects, namely the person who expresses the sentiment (i.e. the holder), to what or whom the sentiment may be expressed towards (i.e. the target) and the nature of sentiment (i.e. "polarity", e.g. either "positive", "negative" or "neutral"). The social media provides a platform to coordinate the nature of sentiment. Social media sites, namely Facebook, Twitter and many more, have experienced an increase in number of users, and this increase creates a big data which might have some interesting textual information on sentiments. The increase can be attributed to what users think about social media; that is, it enables pictures, social blogs, wikis, videos, Internet forums, rating, weblogs, podcasts, social bookmarking and microblogging to be shared and re-shared. There is a huge data on social media which can be analysed to find significant and relevant patterns for the benefit of business. Twitter and Facebook, in some countries, are mostly preferred "social media and social networking sites" (Kaplan and Haenlein 2010) which provide corporate communication practitioners with big data that can be mined and to analyse corporate engagement with its stakeholders (Bellucci and Manetti 2017). These sites represent a public platform for expressing opinions of corporate citizenship and stakeholder interests where sentiments are presented and debated (Whelan et al. 2013). Hence, in this chapter, we present the methods and techniques for extracting sentiments that are shared on social media sites.

2 Text Mining

Text mining generates useful information and patterns about people's interaction. Mostly, people share their opinion and fact on issues via the social media using text (Liu 2010). The importance of text mining is that it helps to obtain objective and subjective views on people. Mostly, facts are based on objective expression while opinions are subjective expressions by people. Kasture and Bhilare 2015 opines that algorithms can play an important role in extracting objective and subjective expressions from large amount of data with minimal effort. Algorithms helps to automate the extraction process instead of using manual process of text extraction. The challenge with textual data includes inconsistent text, frequently changed context and usage of text. Addressing the challenge of text mining algorithm helps organisations to find meaningful insights about social network posts that are generated from different users across the globe. It also enables organisations to predict behaviour of customer. In view of this, the "frequency" and the time of posts, which express a thought or opinion, play a key role in text mining. Generally, approach for text mining finds trends that points to either positive, negative or neural feeling expressed by users. In this regard, algorithms on text mining should be well adapted to discover these trends when large amount of data is involved.

- ***Process for text mining***

The process is based on the use of natural language processing technology which applies computational linguistics concepts to interpret text data. This process starts with categorising, clustering and tagging of text. Afterwards, text data is summarised to create a taxonomy of text. Finally, extract information on the text in terms of frequencies and relationship between texts. Algorithms that help with text analysis which are based on natural language processing are based on statistical- or rule-based models.

The process for text mining has been applied to analysis data from social media sites. For instance, an empirical study was conducted on microblog data (such as Twitter) of user sentiments on manufacturing products in order to get user reaction and reply (Agarwal et al. 2011). The study developed a model that classifies tweets into positive, negative and neutral in order to detect and summarise overall sentiments. This model is a two-way classification step. The first step classifies sentiments as either positive or negative classes. The second step classifies sentiments as a three-way task such as "positive, negative and neutral" classes. The model was able to get accurate user reactions and replies on microblogs.

3 Classification Levels of Sentiment Analysis

Basically, the "classification level of sentiment analysis" is performed at three levels, namely document, sentences and aspect levels. The document level is when the whole document expresses a thought that could either be positive or be negative (e.g. product or movie review). The sentence level is when the sentence expresses a "positive, negative or neutral" thought (e.g. news sentence). In this context, the sentence could be either subjective thought (i.e. conjecture) or objective thought (i.e. factual information) (Devika et al. 2016). The aspect level is when text are broken down into different aspects such as attributes and each attribute is allocated to a particular sentiment.

With the current dispensation of big data, more advance approach to discover these levels of sentiment and extract the necessary pattern on sentiment is significant, as this helps to address challenges of inconsistent categorising of text, tagging and summarising of text data. Deep learning is one of such emerging method to sentiment analysis, which has attracted much attention of most researchers. The method, approach and empirical studies on deep learning models are discussed in this chapter.

4 Aspects of Sentiments

The aspect of sentiments is namely holder, target and polarity. The technique for detecting these aspects is discussed as follows:

- ***Holder detection***

The holder represents someone who has an opinion or the source of opinion. The view source identification is an information extraction task that is achieved based on "sequence tagging and pattern matching techniques". The linear-chain CRF model is one of the models that help to identify source of opinion in order to extract the patterns in the form of features (such as part of speech, opinion lexicon features, semantic class of words, e.g. organisation or person) (Wei n.d). For instance, given a sentence x, in order to find the label sequences y, the following equation is used:

$$P(y|x) = \frac{1}{Z_x} \exp\left(\sum_{i,k} \lambda_k f_k(y_{i-1}, y_i, x) + \sum_{i,k} \lambda'_k f'_k(y_i, x) \right) \quad (1)$$

where y_i is expressed as ("S","T","–"), λ_k and λ'_k are defined parameters, f_k and f'_k are feature functions, and Z_x is the factor of normalisation. Thus, given this sentence:

International	officers	believe	that	the	EU	will	prevail
S	T	-	-	-	-	-	-

Extraction Pattern Learning: It computes probability of pattern being extracted at the source of opinion. This probability is expressed by:

$$P(\text{source}|\text{pattern}_i) = \frac{\text{correct sources}}{\text{correct sources} + \text{incorrect sources}} \quad (2)$$

Thus, the pattern that was extracted on features is expressed using four IE pattern-based features for each token x, that is, *SourcePatt-Freq*, *SourcePatt-Prob* and *SourceExtr-Freq*, *SourceExtr-Prob*, where *SourcePatt* shows whether a word "activates any source extraction pattern", e.g. "complained" activates the pattern "complained", and where "*SourceExtr* indicates whether a word is extracted by any source pattern", e.g. "They" would be extracted by the "complained".

- ***Target identification/detection***

This is what or whom the sentiment is expressed to (i.e. the target). For instance, "customer reviews" of a product/brand name. In order to carry out the review, it is important to identify the features of products. One of the method to achieve this is association rule mining (Hu and Liu 2004).

- ***Polarity detection***

This is the process of mining in order to summarise reviews such as customer reviews, using either positive, negative or neutral opinion of customers so as to find mostly prevailed opinion (Hu and Liu 2004).

5 Sentiment Analysis Framework

Sentiment analysis can be a classification problem that can be grouped into the following stages:

5.1 Review Stage

The source of data for "sentiment analysis" process is from product review from people. Since people are the main source of opinions and emotions on product review, these reviews include news articles, political debates and many more.

5.2 Sentiment Identification Stage

Identification stage is when the specific sentiment is identified, in the form of words on reviews or phrases.

5.3 Feature Selection Stage

This stage ensures the extraction and selection of text features. Some to these features include terms presence and "frequency", n-grams (i.e. n items a given sequence of text), part of speech, opinion words and phrases, and negations (Medhat et al. 2014). Terms presence and "frequency" are the individual words and their term counts. Mostly, selection techniques use bagging of words, that is, when similar words are grouped together. This helps to create a taxonomy of words to indicate the relative importance. Part of speech refers to words used to express opinions or phrases that expresses opinions without using opinion words. Negations refer to the use of negative words which "may change opinion orientation like not good" which is equivalent to "bad".

The methods for feature selection include the lexical-based and statistical-based methods. The lexical-based method allows human annotation, while statistical method applies statistical techniques to automate the selection process. Statistical methods include pointwise mutual information, chi-square and "latent semantic indexing" (Medhat et al. 2014).

5.4 *Sentiment Classification Stage*

This stage presents the approach for classification of sentiments as follows:

5.4.1 Formal Grammar Approach

Formal grammar approach is one of the approaches to sentiment analysis (Kasture and Bhilare 2015) which uses linguistic features. This approach considers the syntactic of text and extracts the different structures of text-like sentences, phrases and words, in order to find a binary relation for dependency grammar. The actual sentiment analysis can be performed by classifying the user sentiments as "positive" or "negative". The impact of each sentiment is then compared with the "subject" and "object" of the sentence. Formal grammar approach relies of structure of text that relates to the lexical structure. Therefore, formal grammar approach also refers to lexicon-based approach as it applies "linguistic features". Thus, "lexicon-based approach relies on sentiment lexicon, which collects known and precompiled sentiment terms". This can be grouped into "dictionary-based and corpus-based" approaches that applies "statistical or semantic" methods to identify "sentiment polarity" (Medhat et al. 2014).

The advantage of formal grammar approach is precision in assigning polarity value on lexical level, thus guaranteeing sentiment classification. Secondly, it can be applied to any domain. However, the disadvantage is the robustness in the sense that if there are missing or incorrect axioms, the system will not work as desired (Kasture and Bhilare 2015).

Machine learning approaches to sentiment analysis are described and discussed as follows:

5.4.2 Machine Learning Approach

The machine learning approach trains dataset to predict outcome on sentiment. Machine learning approaches can be classified in supervised, semi-supervised and unsupervised learning. Supervised learning is when features are labelled and an algorithm predicts come from input feature; semi-supervised is applied when some features are labelled but most of it is not labelled which requires the use of an algorithm to predict an outcome, while unsupervised is when features are unlabelled and algorithm predicts the outcome.

The labelled training set is the input feature vector with corresponding class labels. On the other hand, the test set is used to validate a model's prediction of a class label of unseen feature. Machine learning techniques are namely "naïve Bayes", "maximum entropy", "support vector machine" (SVM) and "deep learning".

Supervised learning method is applied when classes of features that express opinions are labelled for training. In order to learn from labelled classes, an algorithm is applied to learn from training data to predict an outcome. The following are some of the supervised learning methods.

Naïve Bayes Method

This method applies conditional probability to classification of features and mostly applied if the size of training set is small. The method is based on Bayes theorem in which the conditional probability that an event X occurs given the evidence Y is determined by Bayes rule. This rule is expressed as:

$$P(X|Y) = P(X)P(Y/X)/P(Y) \tag{3}$$

where X represents an event and Y is evidence. The equation is expressed in terms of sentiment and sentence as:

$$P(\text{Sentiment/Sentence}) = P(\text{Sentiment})P(\text{Sentence/Sentiment})/P(\text{Sentence}) \tag{5}$$

The advantage of naïve Bayes is that it is simple and intuitive method. It combines efficiency with reasonable accuracy. The disadvantage is that it cannot be used on large dataset. It assumes conditional independence among the linguistic features.

Maximum Entropy Classifier Method

This method applies set of weighting values to combine the joint features that are generated from a set of features. In the process of joining features, it encodes each feature and then maps a related feature set and labels to a vector. The maximum entropy classifier works by "extracting set of features from the input, combining them linearly and then using the sum as exponent". If this method is done in an unsupervised manner, then "pointwise mutual information (PMI) is used to find the co-occurrence of a word with positive and negative words" (Devika et al. 2016). The advantage is that it does not assume the "independent features" as in naïve Bayes method. The disadvantage is that it is not simple to use.

Support Vector Machine

This method is a "machine learning algorithm for classification problems". The method is useful in text and hypertext categorisation. This method does not use probability; instead, it makes use of decision planes to define decision boundaries. The decision plane helps to separate set of features into class, and each class is separated by a separating line. SVM finds a hyperplane with largest possible margin (Devika et al. 2016). SVM requires training set and use of kernel for "mapping" or

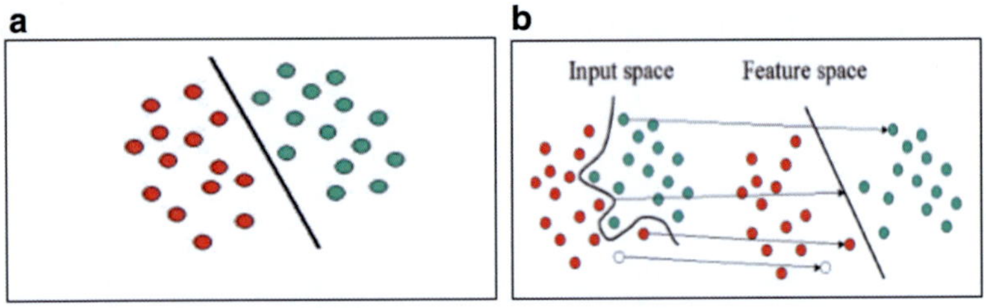

Fig. 1 **a** Linear classifier. **b** SVM illustration

"transformation". After transformation, the mapped features are "linearly separable, and as a result the complex structures having curves to separate the classes can be avoided".

The advantage of SVM is high-dimensional input space. This requires setting "few irrelevant features" and documents vectors sparsely represented. The disadvantage is a huge amount of data training set is required.

The advantage of using machine learning approach is that there is no need to create dictionary of words. It also leads to high accuracy of classification. The disadvantage is training classifiers on text in one domain which in most cases does not work with other domain.

Most research into sentiment analysis uses Twitter messages because of the huge amount of data it generates. Companies have taken advantage of this huge number of users to market their products. The existing literature on sentiment analysis from Twitter dataset used various "feature sets and methods, many of which are adapted from more traditional text classification problems" (Ghiassi et al. 2013). Thus, feature set reduction should be considered in feature classification problems. An approach to supervised feature reduction includes "n-grams and statistical analysis approach" to create a "Twitter-specific lexicon for sentiment analysis" that is brand-specific.

Weblog is one of the ways by which users share their opinions on topics on the World Wide Web (Durant and Smith 2006) with virtual communities. Usually, when a reader turns to "weblogs as a source of information, automatic techniques identify the sentiment of weblog posts" in order to "categorise and filter information sources". However, sentiment classification of political weblog posts appears to be a more difficult classification problem because of the "interplay among the images, hyperlinks, the style of writing and language used within weblogs". Naïve Bayes classifier and SVM are approaches used to predict category of weblog posts.

Empirical review on document-level sentiment analysis uses movie reviews, and combined naïve Bayes and linear SVM to build a model (see Fig. 1) to analyse sentiments on heterogeneous features (Bandana 2018). In this model, heterogeneous features were created based on "combination of lexicon (like SentiWordNet and WordNet) and machine learning (like bag-of-words and TF-IDF)". The approach was applied to movie review in order to classify text movie reviews into polarity such as positive or negative.

The proposed model consists of five components, namely the movie review dataset, pre-processing, feature selection and extraction, classification algorithms and sentiment polarity (Bandana 2018). The process is summarised as follows: an input, manually created movie review text documents are used, but when they are collected from the Web there are irrelevant and unprocessed data that must be pre-processed using different data pre-processing techniques other than "feature selection and extraction". After getting a feature matrix, the matrix is applied on different supervised learning classifiers such as linear support vector machines and naive Bayes, which can be used to predict sentiment label to give a reviewed text polarity orientation either as positive or as negative (Bandana 2018).

The challenge of the proposed approach for heterogeneous feature is that it is not suited for large data processing and this could affect the accuracy of sentiment analysis. Hence "deep learning features such as Word2vec, Doc2Paragraph and word embedding apply to deep learning algorithms such as recursive neural network (RNN), recurrent neural networks (RNNs) and convolutional deep neural networks (CNNs)" that could improve on the accuracy of sentiment analysis from heterogeneous features and guarantee remarkable result (Bandana 2018).

Deep Learning Methods

This method for classification is machine learning approach that is applied for sentiment analysis (Zhang et al. 2018). Conceptually, deep learning uses "multiple layers of nonlinear processing units for feature extraction and transformation". Because multiple layers are used in deep learning, it needs large amount of data to work with (Araque et al. 2017). Thus, using it for feature extraction requires large number of features to be fed into the deep neural network. When these features are in the form of words, then large words are fed into models built using deep learning concept. The words could relate to user sentiment. As sentiments are collected, it needs to be transformed to make discovering of opinions clearer, useful and accurate. The transformation process uses mathematical models to map words to real numbers. Thus, "deep learning" for sentiment analysis needs word embedding as input features (Zhang et al. 2018). The word embedding is a technique for language modelling and feature learning, which transform "words in vocabulary to vectors of continuous real numbers". Generally, deep learning approach starts with input sentences or features in a sequence of work. In this context, each word represents a one vector and each subsequence word is projected into a "continuous vector space by being multiplied with a weighted matric that forms a sequence of real value dense" (Hassan and Mahmood 2017).

Empirically, deep learning models have been applied to model inter-subjectivity problems in sentiment analysis. Inter-subjectivity uses convoluted neural network to find the gap between surface form of a language and corresponding abstract concepts (Gui et al. 2016). The advantage of deep learning models is that it requires large amount of training data in order to make prediction of sentiment. Deep learning has also been applied in the financial sector to predict volatility using financial disclosure sentiment with word embedded-based information retrieval models (Rekabsaz et al. 2017). Deep learning has also been applied in opinion recommendation in which a customised review score of a product from a user (Wang and Zhang 2017). Deep learning has been applied for stance detection with bidirectional LSTM with conditional encoding to detect stance in political Twitter data (Augenstein et al. 2016).

Deep learning methods have been used to identify specific location (using the z-order, that is a spatial indexing method) of users and their text on one site and the corresponding review available on different social media sites. Preethi et al. (2017) present recurrent neural network (RNN) based on deep learning system for sentiment analysis of short text and corresponding reviews available on different social media sites. This approach analyses different reviews, computes an optimised review score and consequently recommends an analysis to each user. The RNN based on deep learning model is used to process sentiments collected from different sites connected with each other so as to classify review as positive and negative (Preethi et al. 2017).

Empirical study by Stojanovski et al. (2015) applied the deep CNN architecture for sentiment analysis on Twitter message. Their approach applied multiple filters and nonlinear layers placed on top of the convolutional layer to ensure the classification of Twitter messages.

Liao et al. (2017) analysed sentiment in Twitter data and suggested a model for predicting user satisfaction of products considering their feeling of a particular environment. The model suggested was based on simple CNN as it classifies feature from global information piece by piece and finds relationships among features in the form of text data. The model was tested with two datasets, namely MR- and STS-Gold datasets. MR dataset consists of a "set of movie reviews with one sentence per review, and the reviews are from Internet users and are similar to Twitter data". STS-Gold Dataset consists of real Twitter dataset. After training of the CNN with datasets, the Twitter data was inputted using "hashtag and stored in MongoDB". The convolutional neural network then outputs the sentiment as positive and negative. The challenge with the model is that could not consider location data on where review emanated from and could not consider multimedia data. It is possible that when large data such as "spatial data of geo-tag" and multimedia data is required performance will be challenged; therefore, other methods such as deep CNN can help resolve this challenge.

The neural network has been applied in many research works such as for document recognition task (Chu and Roy 2017), analysing both the visual and audio for image classification (Krizhevsky et al. 2012) and speech recognition (Chu and Roy 2017). However, the era of big data has made document recognition, and audio and visual analysis for sentiments a difficult task. This is because all features are not labelled. Hence, deep learning method can also be applied when feature is not labelled (i.e. unsupervised that was used for pattern analysis) and when some features are labelled but most of it is unlabelled (i.e. semi-labelled). LeCun et al. (2015) indicate that unsupervised learning such as deep learning has catalytic effect to overshadow the supervised learning. This is because deep learning requires very limited "engineering by hand, so it can easily take advantage of increases in the amount of available data" such as social media data (LeCun et al. 2015). The advantage of catalytic effect on trend of pattern analysis on social media is that it will help to explore fatigue-related issues related to driving and avoid road accidents. Empirical study conducted by Chu and Roy (2017) on the use of "deep convolutional neural network was based on AlexNet Architecture for the classification of images and audio". This shows the potential of deep learning neural methods for social media images and audio pattern analyses to detect sentiments.

Detecting emotions from images which also refers to affect analysis helps to recognise emotions by semiotic modality. Affect analysis model basically consists of five stages, namely "symbolic cue, syntactical structure, word-level, phrase-level and sentence-level analysis" (Medhat et al. 2014). Affect emotion words can be identified using corpus-based technique. This technique finds opinion words within context-specific orientation either positive or negative. The method finds pattern that occurs together such that "a seed list of opinion words" link other opinion words in a large corpus. The link is connectives like "AND, OR, BUT, EITHER-OR" (Medhat et al. 2014). The challenge with the corpus-based is that it requires a lot of human effort to prepare large corpus on words and it also requires domain expert to create the corpus. In this regard, statistical approaches are applied to find the co-relationship between each opinion word. This avoids the situation of unavailability of some words in large corpus. Thus, polarity of a word is determined by the frequency of occurrence of word. In this context, words with similar frequency appear together in a corpus to form a pattern. The combination of affect analysis models with deep learning methods presents unique opportunity for sentiment analysis models because of large data available on social media.

Vosoughi et al. (2015) looked at whether there is a correlation in different locations, times and authors with different emotional valences. The approach applied distant technique to gather labelled tweets from different locations, times and authors. Afterwards, variation of tweet sentiments across diverse authors, times and their locations was analysed to understand the relationship between variables and sentiment. In this study, Bayesian methods were applied to combine different variables with "standard linguistic features", namely "n-grams" to create a "Twitter sentiment classifier". Thus, integrating a contextual information seen on Twitter into "sentiment classification" problem is a very promising research area (Vosoughi et al. 2015). The concept of deep learning may be explored within this context as well.

Sun et al. (2018) analysed the sentiment in the "Tibetan language". Tibetan is an independent language and writing system for Tibetans. Apart from China, the language is spoken by people in Europe, Nepal, Bhutan and India. It was estimated that 7.5 million people around the world used Tibetan at the end of 2017 (Sun et al. 2018). It is common for Tibetans to express their opinions and emotions on social media, and based on this, a multi-level network was built based on deep learning model for the classification of emotional features from the "Tibetan microblogs" in order to find sentiment that describes emotions expressed using Tibetan. Tibetan word microblog was used to test the model. At the initial stages, the model was trained as a word vector by using the word vector tool; then, the trained word vectors and the corresponding sentiment orientation labels are directly introduced into the different deep learning models to classify the Tibetan microblogs.

Shalini et al. (2018) applied CNN to classify sentiments in India. India was selected because of the diversity of language it uses. Basically, the Bengali–English is mainly spoken; in view of this, it resulted in the evolution of code-mixed data, which is a combination of more than one language. The convolutional neural network was applied to develop a model for the classification of sentiments into positive, negative or neutral. Initially, an input word vector of n dimension corresponding to word in the sentence is inputted into the model. The convolution operation is done on the input sentence matrix using a filter. These filters undergo "convolution by sliding the filter window along the entire matrix". The output of each filter then "undergoes pooling which is done using max operation". Moreover, the "pooling techniques help in fixing the size of the output vector and also help in dimensionality reduction". The pooling layer output is fed to the "fully connected *Softmax* layer where the probability of each label is determined" (Shalini et al. 2018). The "dropout regularisation technique is used to overcome over-fitting". While training, it is done by removing some of the randomly selected neurons. However, better accuracy for code-mixed data can be achieved by using *Word2vec* instead of word indexing.

Similarly, Ouyang et al. (2015) present a model that combines *Word2vec* and convolutional neural network (CNN) for sentiment analysis on social media. Initially, the Word2vec computes vector representations of words which is fed into the CNN. The basis of using *Word2vec* is to find the vector representation of word and to determine the distance of words. In view of finding the distance, parameters were initialised so as to find good point of CNN and improve on performance. The model architecture applied 3 pairs of convolutional layers and pooling layers, "parametric rectified linear unit (PReLU), normalisation and dropout technology to improve the accuracy and generalisability of the model". The model was validated using dataset including "corpus of movie review excerpts that includes five labels: negative, somewhat negative, neural, somewhat positive and positive".

Alshari et al. (2018) created lexical dictionary for sentiment analysis. The approach used SentiWordNet, which is the most used sentiment lexical to "determine the polarity of texts". However, a huge number of terms in the corpus vocabulary are not in the SentiWordNet because of the "curse of dimensionality" and this reduces the "performance of the sentiment analysis". In order to address this challenge, a method was proposed to help enlarge the dictionary by learning the polarity non-opinion words in the vocabulary based on the SentiWordNet. The model was evaluated on Internet Movie Review Dataset. The proposed Senti2Vec method was more effective than the SentiWordNet as the sentiment lexical resource (Alshari et al. 2018).

Deep learning plays an important role in "natural language processing" as far as the use of "distributed word representation" is concerned. Real-value vector representation in "natural language processing" finds similarity between words and concepts. The hierarchical structure of deep learning architecture has helped to create a parallel distributed processing which is much useful for large word processing. In view of this, deep learning has enormous potential. However, deep learning models should be self-adaptive to minimise the error in prediction during the real-value mapping of words and concepts. In this regard, random search algorithms are significant for self-tuning of deep learning models. Hence, bio-inspired approaches could help to achieve this self-tuning of parameters in deep learning models.

Bio-inspired or meta-heuristic-based algorithms are emerging as an approach to sentiment analysis. This is because it helps to select optimal subset of features and eliminate features that are irrelevant to the context of analysis, thereby enhancing performance of classification to guarantee accurate results.

5.4.3 Bio-inspired Methods

Practitioners and academics have developed models based on bio-inspired algorithms to facilitate data analytics for big data. These bio-inspired algorithms can be categorised into three domains: ecological, swarm-based and evolutionary (Gill and Buyya 2018). Swarm and evolutionary-based algorithms are inspired by the collective behaviour and natural evolution in animals, while ecology-based algorithms are inspired by ecosystems which involves an interaction of living organisms in their abiotic environment such as water, soil and air (Binitha and Sathya 2012). Ecology-inspired optimisation is one of the most recently developed groups of bio-inspired optimisation algorithms (Čech et al. 2014), although the most commonly used and researched optimisation methods are evolution-inspired algorithms which use the principle of evolution and genetics to address prevailing problems in big data analytics. Swarm intelligence-based algorithms are the second well-known branch of biology-inspired optimisation (ibid).

Genetic Algorithm

Ahmad et al. (2015) proposed a model for feature selection in sentiment analysis based on natural language processing, and genetic algorithm and rough set theory. The document dataset was used to test this model. Initially, this model extracts sentences from the documents and performs data pre-processing by removing stop-words, stemming, misspelled words and part-of-speech (POS) tagging, thereby improving the quality of analysis. In POS tagging, a sentence is parsed and respectively the features are identified and extracted. Finally, meta-heuristic algorithm is used for selecting the set of optimum features.

Ant Colony Optimisation (ACO) Algorithm

Ant colony optimisation which forms part of the swarm intelligence was applied for opinion mining on social media sites. ACO is based on the behaviour of ants and how they find food sources and its home. The approach started by data collecting from Twitter in the form of list JSON which have various attributes with all the information about the post like the number of retweets (Goel and Prakash 2016). Twitter data has the following format: "User_id", "id_str", "created_at", "favourite_count", "retweet_count", "followers_count", "text"; the value and definition of all these can be found at the information page of Twitter's tweepy API. Data from Reddit was collected using Praw API. The data is pre-processed and normalised. During pre-processing, posts are tokenised into different words, and links, citations, etc., are also removed as they do not convey sentiment. Words are also stemmed to their root words so as to make it easier to classify them as conveying positive or negative sentiment. The swarm algorithm was then applied, in which the evaporation (of opinion) emphasises the path preferred (positive/negative) by the users in our conversation (Goel and Prakash 2016). The paths which have been trained are used to evaluate the learning of the algorithm. The ants make the prediction according to the weights (heuristic), and then the sentiment is evaluated for the post. Whenever a prediction does not match the sentiment, the error value is incremented. This is only done in the testing phase where the tenth "subset" of the data is used. The results indicate that, in the case of the Twitter dataset, 1998 records were selected for testing after the remaining records were used in the training part to get a list of the required values to be checked for the build-up of opinion. Of these 1998, 1799 records were correctly predicted by the algorithm and 199 were incorrectly predicted. This gives us an accuracy of 90.04%. Whereas for the Reddit dataset 289 records were selected for testing and the remaining dataset was used for training. Two hundred and ten records resulted in correct sentiment prediction, while 79 resulted in incorrect prediction leading to an accuracy of 72.66%. Based on the results from Twitter and Reddit datasets, the accuracy is lower in Reddit dataset because of number of records in the dataset and also there are vast differences in lengths of the various posts. However, these results on accuracy can be improved by use of robust natural language techniques. Another challenge with this approach is that algorithm does not perform well when sentiment changes quickly and drastically in group chats (Goel and Prakash 2016).

Redmond et al. (2017) present a tweet sentiment classification framework which pre-processes information from emoticon and emoji. The framework allows textual representation in tweets, and once tweets are pre-processed, a hybrid computational intelligence approach classifies the tweets into positive and negative. The framework combines three methods, namely the "singular value decomposition and dimensionality reduction method to reduce the dimensionality" of the dataset; the "extended binary cuckoo search algorithm, to further reduce the matrix by selecting the most suitable dimensions; and the support vector machine classifier which is trained to identify whether a tweet is positive or negative". During the experiment to validate the model, total of "1108 tweets were manually extracted, and each tweet was assigned a class value of 1 for positive or 0 for negative; thus, a total of 616 were positive and 492 were negative tweets". The technique to evaluate the model's performance measures is "precision", "recall" and the "F1-measure". The experimental results show that the proposed approach yields "higher classification accuracy and faster processing times than the baseline model which involves applying the extended binary cuckoo search algorithm and support vector machine classifier to the original matrix, without using singular value decomposition and dimensionality reduction" (Redmond et al. 2017). The challenge with this model is that it is not adapted to the finding multiple classes of tweets. Additionally, since this model was not applied to large dataset, the accuracy of classification may be challenged. Hence, new models can be built to enable large sentiment analysis and to find multiclass classification of tweets.

Hybrid Particle Swarm Optimisation (PSO) and Ant Colony Optimisation (ACO) Algorithm

Stylios et al. (2014a) conducted a study to extract users' opinions from text Web sources (e.g. blogs) and classified posting into two, namely post supported by argument and post not supported by argument. The study applied bio-inspired algorithm, namely the hybrid PSO/ACO2 algorithm to classify Web posts in tenfold cross-validation experiment. The technique extracts user's opinion from real Web content textual data on product information. Initially, the approach collected "content of the users' posts and non-textual elements (images, symbols, etc.) are eliminated by applying HTML". Secondly, "tokenisation is applied to the postings' body to extract the lexical elements of the user-generated text". Afterwards, the "text is passed through a part-of-speech tagger, responsible for identifying tokens, and annotates them to appropriate grammar categories". Additionally, the topics of discussion and the user's opinion on the topics were identified. In order to detect such references within a post, a "syntactic dependency parser is applied to identify the proper noun to which every adjective refers to when given as input text containing adjectives". Similarly, to identify opinion phrases on user's postings, the "adjective–noun" pairs are used so as to build a dataset on opinion. In order to "detect how users assess commercial products, the motion of word's semantic orientation is used". To obtain the "semantic frame of an adjective, every adjective extracted from the harvested postings against an ontology which contains fully annotated lexical units is examined". Sentiment analysis of "users' opinions refers to labelling opinion phrases with a suitable polarity tag (positive or negative) to the adjectives". The criterion under which "labelling takes place is that positive adjectives give praise to the topic, while negative adjective criticises it". Finally, the model was validated using a database which consists of the "extracted features as well as the annotation per post provided by an expert, for a total of 563 posts". The classification schema used consists of the PSO/ACO2 and C4.5 algorithms, trained and tested with the database. The "PSO/ACO2 is a hybrid algorithm for classification rule generation". The rule discovery process in PSO/ACO2 algorithm is performed into two separate phases. Specifically, "in the first phase, a rule is discovered using nominal attributes only, using a combination of ACO and PSO approach". In the second phase, the rule is extended with continuous attributes. "The PSO/ACO2 algorithm uses a sequential covering approach to extract one classification rule at each iteration". The bio-inspired algorithm PSO/ACO2 was "superior classification performance in terms of sensitivity (81,77%), specificity (94,76%) and accuracy (90.59%), while C4.5 algorithms produce classification performance results in terms of sensitivity (73.46%),

Table 1 Swarm intelligence techniques on sentiment analysis

Swarm intelligence technique	Author and year	Dataset	Classifier	Accuracy without optimisation	Accuracy with optimisation
ABC	(Dhurve and Seth 2015)	Product reviews	SVM	55	70
ABC	(Sumathi et al 2014)	Internet Movie Database (IMDb)	Naïve Bayes	85.25	88.5
			FURIA	76	78.5
			RIDOR	92.25	93.75
Hybrid PSO/ACO2	(Stylios et al. 2014b)	Product reviews and governmental decision data	Decision tree	83.66	90.59
PSO	(Hasan et al. 2012)	Twitter data	SVM	71.87	77
PSO	(Gupta et al. 2015)	Restaurant review data	Conditional random field (CRF)	77.42	78.48

Source (Kumar et al. 2016)

specificity (87,78%) and accuracy (83.66%)" (Stylios et al. 2014a). The significance of this study is that it presents unique potential of bio-inspired algorithms to sentiment analysis.

Table 1 presents comparison of swarm intelligence techniques on sentiment analysis and the accuracy as follows:

Table 1 shows the application of swarm intelligence algorithm for sentiment analysis. Bio-inspired algorithms have been applied on sentiment analysis to improve the accuracy of classification. However, with the current dispensation of big data, the accuracy of classification algorithms may be challenged as learning algorithm explore different parameters to find the best or near best parameter which can produce better classification result. Moreover, Sun et al. (2018) indicate that accuracy can be improved when different optimisation parameters are applied to sentiment analysis models. Bio-inspired algorithms can help to find optimal parameter in a classification problem and although, bio-inspired algorithms for classification have been proposed, not a number of these bio-inspired algorithms have been combined with deep learning methods for classification of sentiments. Such algorithm is kestrel-based search algorithm (KSA) which combines deep learning method (i.e. recurrent neural network with "long short-term memory" network) (Agbehadji et al. 2018) for general classification problem. The results show that KSA is comparable to BAT, ACO and PSO as the test statistics (i.e. Wilcoxon signed-rank test) show no statistically significant differences between the means of classification accuracy at level of significance of 0.05. Thus, KSA shows some potential for sentiment analysis.

In summary, the sentiment classification methods can be presented as follows.

1. Machine learning approach:
    (a) Supervised learning:
        (i) Decision tree classifiers
        (ii) Linear classifiers, namely support vector machine and neural network
        (iii) Rule-based classifiers
        (iv) Probabilistic classifiers, namely naïve Bayes, Bayesian network and maximum entropy
    (b) Unsupervised learning
2. Lexicon-based approach:
    (a) Dictionary-based approach
    (b) Corpus-based approach, namely statistical and semantic.

5.5 Polarity Stage

This stage predicts the sentiment class as either positive, negative or neutral. Machine learning such as "naive Bayes classification", "maximum entropy classification" and SVM as discussed earlier was some of the methods used to find polarity. For instance, all WordNet synsets were automatically annotated for degrees of positivity, negativity and "neutrality/objectiveness" (Baccianella et al. 2010).

Raghuwanshi and Pawar (2017) present a model on both sentiment analysis and sentiment polarity categorisation of online Twitter dataset. In this model, the SVM and two probabilistic methods (i.e. logistic regression model and naïve Bayesian classifier) were applied. Initially, tweets dataset was loaded from Twitter.com to test the model. Afterwards, the following steps are applied:

a. Tokenising—splitting sentences and words from the body of text
b. Part of speech tagging
c. Machine learning with algorithms and classifiers
d. Tie in scikit-learn (sklearn)
e. Training classifiers with dataset
f. Performing live, streaming, sentiment analysis with Twitter.

During the experiment, a tenfold cross-validation was applied as follows: the dataset is partitioned into 10 equal size subsets, each of which consists of 10 positive class vectors and 10 negative class vectors (Raghuwanshi and Pawar 2017). One of the 10 subsets are selected, and that 1 subset is maintained as validation dataset to test a classification model with others, whereby remaining 9 subsets are used as training dataset. Performance of each classification model is estimated by generating confusion metric with the calculation and comparison of results on precision value, recall value and F1-score. The accuracy of each algorithm, namely SVM, logistic regression model and naïve Bayesian classifier, is 78.82%, 76.18% and 71.54%, respectively. The results show that SVM gives high accuracy value.

6 Sentiment Analysis on Social Media Business Communication

The transition from a monolithic to a dialogue-based approach to business communication has popularised the use of data mining, natural language processing (NLP), machine learning ethnographic data analysis techniques in the analysis of trends and patterns in consumer behaviour, especially in relation to discharging holistic corporate citizenship duties (Crane and Matten 2008). While corporate citizenship duties such as corporate social responsibility (CSR) are gaining popularity in emerging economies (KPMG 2013), underlying motivations behind CSR efforts have generated more controversy in recent years, which has increased levels of corporate legitimation disclosure (Cho et al. 2012) and crisis communication (Coombs and Holladay 2015) on social media.

This phenomenon has created more need for algorithm-based corporate communication techniques, especially on social media where opinions are freely expressed. Discrepancy between expectations and actual corporate behaviour on CSR has been addressed based on how relevant publics respond to CSR disclosure strategies (Merkl-Davies and Brennan 2017) and contents of disclosures on social media (Gomez-Carrasco and Michelon 2017), the level of engagement in comments (Bellucci and Manetti 2017), stakeholders' reactions to CSR disclosure strategies—by exploiting big data about the interactions between firms and stakeholders in social media (She and Michelon 2018).

Carroll (1991) identified four main expectations of CSR: economic (profitability), legal (compliance with the law), ethical (conducts) and philanthropic expectations of companies' responsibilities to the society, which are possible domains in which consumers' CSR-related sentiments can be classified and assessed using supervised, semi-supervised and unsupervised algorithm-based sentiment analyses.

Facebook and Twitter are the most popular "social media and social networking sites" (Kaplan and Haenlein 2010) which provide corporate communication practitioners with big data with which to mine and analyse corporate engagement with stakeholders (Bellucci and Manetti 2017). These sites represent a public arena for expressing opinions of corporate citizenship as divergent stakeholder interests and sentiments are presented and debated (Whelan et al. 2013).

Ecologically inspired algorithms generally rely on the hierarchical classification of concepts into relevant categories as adapted from the field of science which describes, identifies and classifies organisms in biology, which is equally applicable to business and economics. Ecology-inspired optimisation is gaining popularity in solving complex multi-objective problems (Čech et al. 2014), such as those of the complexities of CSR-related sentiments on social media.

6.1 Approaches to Supervised, Unsupervised and Semi-supervised Sentiment Analyses in CSR-Related Business Communication: Applications in Theory and Practice

Sentiment analysis is usually carried out using corpus data which is often accompanied with supervised techniques such as content analysis as analysis entails the manual coding and classification of texts and corpus data. Semi-supervised and unsupervised techniques are equally instrumental for identifying prevailing cues in sentiment analysis, with minimal human intervention and intuition in the coding and pre-processing of texts to reduce bias and noise in textual data.

6.1.1 Supervised Sentiment Analysis: A Content Analysis-Based Approach

Sentiment analysis has traditionally been assessed using content analysis-based approaches, in combination with other quantitative techniques. For example, She and Michelon (2018) assessed stakeholders' perception of hypocrisy in CSR disclosures on Facebook based on a content analysis on 21,166 posts related to S&P100 firms. A Python script programming language was used to retrieve data from the application programming interface (API) on Facebook. An ad hoc R script was used to perform textual analysis on Facebook which retrieved data from all emotions, comments and texts of comments under comments associated with the posts. Loughran and McDonald's (2011) "bag-of-words" approach was used to categorise texts followed by a manual checking of posts to eliminate misclassifications, manual coding of CSR-related posts and classifications of posts based on predefined guidelines.

6.1.2 Semi-supervised Sentiment Analysis: Sarcasm Identification Algorithm

Semi-supervised sarcasm identification (SASI) algorithm is another approach used to identify sarcastic patterns and classify tweets based on the probability of being sarcastic, excluding URLs and hashtags, un-opinionated words and noisy and biased contents (Davidov et al. 2010). In the study, tweets were manually inspected to compare results from the system's classification and human classification of sarcastic tweets. The keyword-based approach was used where a collection of sentiment holding bag-of-words were assigned a binary sentiment of either positive or negative or positive and negative lexicons (neutral) (Wiebe et al. 2005). Bugeja (2014) offered a Twitter MAT approach, composed of a Web application user interface and classification module which automatically processes, gathers, classifies and filters tweets, including all algorithms, from the Twitter API. Tweets are shown in JSON format which are then processed through GATE for the annotation of sentiment words—based on positive or negative words and adverbs of degree. Twenty individuals were given different tweets to assign sentiments for each, and matching classifications (75% corresponding agreements) were collated and compared to the results obtained from Twitter MAT and existing sentiment database such as the Stanford Twitter Sentiment Test Set, Sentiment Strength Twitter Dataset and STS-Gold Dataset.

The approach to trend analysis and patterns in sustainability reporting based on text mining techniques (Te Liew et al. 2014; Aureli 2017) has gained popularity in recent years. The occurrences of words or phrases in documents and textual data are subsumed in text mining algorithms (Castellanos et al. 2015). Text mining is a computer-aided tool which retrieves information on trends and patterns from big textual data (Fuller et al. 2011) such as annual reports and social media posts. Text mining is therefore foregrounded on algorithms based on data mining, NLP and machine learning techniques (Aureli 2017) and often used to analyse sentiments in CSR-related social media posts (Chae and Park 2018). The application of text mining is mostly common among accounting and business management researchers and practitioners who also use content analysis to complement and strengthen outcomes of unsupervised and semi-supervised approaches to sentiment analyses. Content analyses are also instrumental for uncovering implicit consumer values and behaviours towards CSR corporate disclosures in annual reports and social media sites.

6.1.3 Unsupervised Sentiment Analysis: Structural Topic Modelling

Structural topic modelling (STM) emanated from the "latent Dirichlet allocation (LDA)" model in the topic modelling technique (Chae and Park 2018). Topic modelling is an "unsupervised machine learning-based content" analytics algorithm which focuses on automatically discovering implicit "latent" structure from large text corpora based on a collection of words which contain multiple topics in diverse proportions. This technique helps to discover and classify potentially misleading posts which could contain overlapping keywords cut across other categories. A topic is a list of semantically coherent words which have different weights. Most LDA-based topic models are developed in machine learning communities whose focus is on discovering the overall topics from big text data.

The multidimensionality of business research and database such as CSR, sustainability and corporate governance structure and reports often lead to additional information or metadata which function as covariates in traditional topic modelling techniques. The extensive need for the analysis of covariates in business research makes STM (Roberts et al. 2016) particularly instrumental in business research. STM is a relatively new probabilistic topic model, which incorporates covariates and supplementary information in topic modelling (ibid.).

In a survey of CSR-related trends and topics using STM, model selection and topic modelling were carried out using the R package, after the pre-processing and classification of the corpus data. STM for the study sought to enhance topic discovery, visualisation, correlation and evolution from CSR-related words in Tweets and assigning a relevant label for each topic based on 1.2 m Twitter posts. Topic modelling gives room for a single search query (*k*) per search. Using a mixed method approach, the optimal number of topics was determined by comparing the residuals of the model with a different (*k*) (from 2 to 80) to ascertain model fitness (30–50). Topic coherence and topic exclusivity were compared where models gave low residual values. Co-appearance of words in a corpus with semantically coherent words and non-overlapping words across other topics was deemed cohesive and exclusive, respectively. The optimal performance of the topic model can be ascertained at a specific topic number, in which case topic number 31 was the optimal performance level of the topic model in this study.

7 Conclusion

The sanitisation of noisy contents in microblogging texts is often a major issue in sentiment analysis (Bugeja 2014). Effort is consistently geared towards reducing bias in data generated on social media sites. Sentiment intensifiers and/or noisy data, namely: hashtags, URLs, capitalised words, emoticons, excessive punctuations and special characters, often make it difficult to classify sentiments (ibid). Confidentiality and anonymity of posts also constitute a challenge in the use of social media textual data, especially for academic research purposes. Nevertheless, social media textual data remains an instrumental source of implicit attitudes and perceptions of complex and sensitive issues around CSR. Sentiment analysis contributes to the understanding of halo-removed CSR-related consumer behaviour, a prototype of the consumer-company endorsement (C-C endorsement) offered by Ijabadeniyi (2018). C-C endorsement is premised on how the congruence between halo-removed CSR expectations and company values can predict corporate endorsements and reputation.

Key Terminology and Definitions

Sentiment analysis is the process of extracting feeling, attitudes or emotion of people from communication (either verbal or non-verbal). Sentiment analysis is also referred as opinion mining.

Social media is an instrument used for communication. An example of such instrument is Facebook.

References

Agarwal, A., Xie, B., Vovsha, I., Rambow, O., & Passonneau, R. (2011). Sentiment analysis of Twitter data. In *Association for Computational Linguistics*, pp. 30–38.

Agbehadji, I. E., Millham, R., Fong, S., & Hong, H. -J. (2018). Kestrel-based Search Algorithm (KSA) for parameter tuning unto long short term memory (LSTM) network for feature selection in classification of high-dimensional bioinformatics datasets. In *Proceedings of the Federated Conference on Computer Science and Information Systems*, pp. 15–20.

Ahmad, S. R., Bakar, A. A., & Yaakub, M. R. (2015). Metaheuristic algorithms for feature selection in sentiment analysis. In: *Science and Information Conference*, pp. 222–226.
Alshari, E. M., Azman, A., & Doraisamy, S. (2018). Effective method for sentiment lexical dictionary enrichment based on word2Vec for sentiment analysis. In: *2018 Fourth International Conference on Information Retrieval and Knowledge Management,* pp. 177–181.
Araque, O., Corcuera-Platas, I., Sánchez-Rada, J. F., & Iglesias, C. A. (2017). Enhancing deep learning sentiment analysis with ensemble techniques in social application. *Expert Systems with Applications, 77*(2017), 236–246.
Augenstein, I., Rocktäschel, T., Vlachos, A., & Bontcheva, K. (2016). Stance detection with bidirectional conditional encoding. In *Proceedings of the Conference on Empirical Methods in Natural Language Processing.*
Aureli, S. (2017). A comparison of content analysis usage and text mining in CSR corporate disclosure.
Baccianella, S., Esuli, A., & Sebastiani, F. (2010). Sentiwordnet 3.0: An enhanced lexical resource for sentiment analysis and opinion mining. *LREC-2010.*
Bandana, R. 2018. Sentiment Analysis of Movie Reviews Using Heterogeneous Features. In *2018 2nd International Conference on Electronics, Materials Engineering & Nano-Technology (IEMENTech)*, pp. 1–4.
Bellucci, M., & Manetti, G. (2017). Facebook as a tool for supporting dialogic accounting? Evidence from large philanthropic foundations in the United States. *Accounting, Auditing & Accountability Journal, 30*(4), 874–905.
Binitha, S., & Sathya, S. S. (2012). A survey of bio inspired optimization algorithms. *International Journal of Soft Computing and Engineering, 2*(2), 137–151.
Bugeja, R. (2014). Twitter sentiment analysis for marketing research. University of Malta.
Carroll, A. B. (1991). The pyramid of corporate social responsibility: Toward the moral management of organizational stakeholders. *Business Horizons, 34*(4), 39–48.
Castellanos, A., Parra, C., & Tremblay, M. (2015). Corporate social responsibility reports: Understanding topics via text mining.
Čech, M., Lampa, M., & Vilamová, Š. (2014). Ecology inspired optimization: Survey on recent and possible applications in metallurgy and proposal of taxonomy revision. In Paper presented at the *23rd International Conference on Metallurgy and Materials.* Brno, Czech Republic.
Chae, B., & Park, E. (2018). Corporate social responsibility (CSR): A survey of topics and trends using twitter data and topic modeling. *Sustainability, 10*(7), 2231.
Cho, C. H., Michelon, G., & Patten, D. M. (2012). Impression management in sustainability reports: An empirical investigation of the use of graphs. *Accounting and the Public Interest, 12*(1), 16–37.
Chu, E., & Roy, D. (2017). Audio-Visual sentiment analysis for learning emotional arcs in movies. *MIT Press*, pp. 1–10.
Coombs, T., & Holladay, S. (2015). CSR as crisis risk: expanding how we conceptualize the relationship. *Corporate Communications: An International Journal, 20*(2), 144–162.
Crane, A., & Matten, D. (2008). The emergence of corporate citizenship: Historical development and alternative perspectives. In A. G. Scherer & G. Palazzo (Eds.), *Handbook of research on global corporate citizenship* (pp. 25–49). Cheltenham: Edward Elgar.
Davidov, D., Tsur, O., & Rappoport, A. (2010). Semi-supervised recognition of sarcastic sentences in twitter and amazon. In *Proceedings of the fourteenth conference on computational natural language learning* (pp. 107–116). Association for Computational Linguistics.
Devika, M. D., Sunitha, C., & Amal, G. (2016). Sentiment analysis: A comparative study on different approaches sentiment analysis: a comparative study on different approaches. *Procedia Computer Science., 87*(2016), 44–49.
Dhurve, R., & Seth, M. (2015). Weighted sentiment analysis using artificial bee colony algorithm. *International Journal of Science and Research (IJSR)*, ISSN (Online): 2319–7064.
Durant, K. T., & Smith, M. D. (2006). Mining sentiment classification from political web logs. In *Proceedings of Workshop on Web Mining and Web Usage Analysis of the 12 th ACM SIGKDD International Conference on Knowledge Discovery and Data Mining*, pp. 1–10.

Fuller, C. M., Biros, D. P., & Delen, D. (2011). An investigation of data and text mining methods for real world deception detection. *Expert Systems with Applications, 38*(7), 8392–8398.

Ghiassi, M., Shinner, J., & Zimbra, D. (2013). Twitter brand sentiment analysis: A hybrid system using n-gram analysis and dynamic artificial neural network. *Journal Expert Systems with Applications, 40*(16), 6266–6282.

Gill, S. S., & Buyya, R. (2018). Bio-inspired algorithms for big data analytics: A survey, taxonomy and open challenges.

Goel, L., & Prakash, A. (2016). Sentiment analysis of online communities using swarm intelligence algorithms. In *2016 8th International Conference on Computational Intelligence and Communication Networks* (pp. 330–335). IEEE.

Gomez-Carrasco, P., & Michelon, G. (2017). The power of stakeholders' voice: The effects of social media activism on stock markets. *Business Strategy and the Environment, 26*(6), 855–872.

Gui, L., Xu, R., He, Y., Lu, Q., & Wei, Z. (2016). Intersubjectivity and sentiment: from language to knowledge. In *Proceedings of the International Joint Conference on Artificial Intelligence (IJCAI 2016).*

Gupta, D. K., Reddy, K. S., Shweta, & Ekbal, A. (2015). PSO-ASent: Feature selection using particle swarm optimization for aspect based sentiment analysis. *Natural Language Processing and Information Systems of the series Lecture Notes in Computer Science, 9103*: 220–233.

Hasan, B. A. S., Hussin, B., GedePramudya, A. I. & Zeniarja, J. (2012). Opinion mining of movie review using hybrid method of support vector machine and particle swarm optimization.

Hassan, A., & Mahmood, A. (2017). Efficient deep learning model for text classification based on recurrent and convolutional layers. *IEEE International Conference on Machine Learning and Applications, 2017,* 1108–1113.

Hu, M., & Liu, B. (2004). Mining and summarizing customer reviews. In *Proceedings of the tenth ACM SIGKDD international conference on Knowledge discovery and data mining.* Seattle, WA, USA.

Ijabadeniyi, A. (2018). Exploring corporate marketing optimisation strategies for the KwaZulu-Natal manufacturing sector: A corporate social responsibility perspective. Ph.D. Thesis, Durban University of Technology.

Kaplan, A. M., & Haenlein, M. (2010). Users of the world, unite! The challenges and opportunities of social media. *Business Horizons, 53*(1), 59–68.

Kasture, N. R. & Bhilare, P. B. (2015). An approach for sentiment analysis on social networking sites. In *International Conference on Computing Communication Control and Automation* (pp. 390–395). IEEE.

KPMG. (2013). *The KPMG survey of corporate responsibility reporting* Available: https://home.kpmg.com/be/en/home/insights/2013/12/kpmg-survey-corporate-responsibility-reporting-2013.html. Accessed 15 Mar 2015.

Krizhevsky, A., Sutskever, I. & Hinton, G. E. (2012). Imagenet classification with deep convolutional neural networks. In *Advances in neural information processing systems*, pp. 1097–1105.

Kumar, A., Khorwal, R. & Chaudhary, S. (2016). A survey on sentiment analysis using swarm intelligence. *Indian Journal of Science and Technology, 9*(39).

LeCun, Y., Bengio, Y., & Hinton, G. (2015). Deep learning. *Nature, 521,* 436–444.

Liao, S., Wang, J., Yua, R., Satob, K., & Chen, Z. (2017). CNN for situations understanding based on sentiment analysis of twitter data. In *8th International Conference on Advances in Information Technology, Procedia Computer Science, 111* (2017), 376–381.

Liu, B. (2007). Web data mining: Exploring hyperlinks, contents, and usage data.

Liu, B. (2010). *Sentiment Analysis and Subjectivity.*

Loughran, T., & McDonald, B. (2011). When is a liability not a liability? Textual analysis, dictionaries, and 10-Ks. *The Journal of Finance, 66*(1), 35–65.

Medhat, W., Hassan, A., & Korashe, H. (2014). Sentiment analysis algorithms and applications: A survey. *Ain Shams Engineering Journal, 2014*(5), 1093–1113.

Merkl-Davies, D. M., & Brennan, N. M. (2017). A theoretical framework of external accounting communication: Research perspectives, traditions, and theories. *Accounting, Auditing & Accountability Journal, 30*(2), 433–469.
Mikalai, T., & Themis, P. (2012). Survey on mining subjective data on the web. *Data Mining and Knowledge Discovery, 2*(24), 478–514.
Ouyang, X., Zhou, P., Li, C. H., & Liu, L. (2015). Sentiment analysis using convolutional neural network. In *IEEE International Conference on Computer and Information Technology; Ubiquitous Computing and Communications; Dependable, Autonomic and Secure Computing; Pervasive Intelligence and Computing*, pp. 2359–2364.
Patel, A., Gheewala, H., & Nagla, L. (2014). Using social big media for customer analytics, pp. 1–6.
Preethi, G., Venkata Krishna, P. V., Obaidat, M. S., Saritha, V., & Yenduri, S. (2017). Application of deep learning to sentiment analysis for recommender system on cloud. In *International Conference on Computer, Information and Telecommunication Systems (CITS)*. IEEE.
Raghuwanshi, A. S., & Pawar, S. K. (2017). Polarity Classification of Twitter data using sentiment analysis. *International Journal on Recent and Innovation Trends in Computing and Communication, 5*(6).
Redmond, M., Salesi, S., & Cosma, G. (2017). A novel approach based on an extended cuckoo search algorithm for the classification of tweets which contain emoticon and emoji. In *2017 2nd International Conference on Knowledge Engineering and Applications*, pp. 13–19.
Rekabsaz, N., Lupu, M., Baklanov, A., Hanbury, A., Dür, A., & Anderson, L. (2017). Volatility prediction using financial disclosures sentiments with word embedding-based IR models. In *Proceedings of the Annual Meeting of the Association for Computational Linguistics (ACL2017)*.
Roberts, M. E., Stewart, B. M., & Airoldi, E. M. (2016). A model of text for experimentation in the social sciences. *Journal of the American Statistical Association, 111*(515), 988–1003.
Shalini, K., Aravind, R., Vineetha, R. C., Aravinda, R. D., Anand, K. M., & Soman, K. P. (2018). Sentiment analysis of Indian languages using convolutional neural networks. In *International Conference on Computer Communication and Informatics (ICCCI-2018)* (pp. 1–4). IEEE.
She, C., & Michelon, G. (2018). Managing stakeholder perceptions: Organized hypocrisy in CSR disclosures on Facebook. *Critical Perspectives on Accounting*.
Stojanovski, D., Strezoski, G., Madjarov, G. & Dimitrovski, I. (2015). Twitter sentiment analysis using deep convolutional neural network. pp. 1–12.
Stylios, G., Katsis, C. D. & Christodoulakis, D. (2014a). Using bio-inspired Intelligence for web opinion mining. *International Journal of Computer Applications (0975–8887), 87*(5), 36–43.
Stylios, G., Katsis, C. D., & Christodoulakis, D. (2014b). Using bio-inspired intelligence for web opinion mining. *International Journal of Computer Applications, 87*(5).
Sumathi, T., Karthik, S., & Marikkannan, M. (2014). Artificial bee colony optimization for feature selection in opinion mining. *Journal of Theoretical and Applied Information Technology, 66*(1).
Sun, B., Tian, F., & Liang, L. (2018). Tibetan micro-blog sentiment analysis based on mixed deep learning. In *International Conference on Audio, Language and Image Processing (ICALIP)* (pp. 109–112). IEEE.
Te Liew, W., Adhitya, A., & Srinivasan, R. (2014). Sustainability trends in the process industries: A text mining-based analysis. *Computers in Industry, 65*(3), 393–400.
Vosoughi, S., Zhou, H., & Roy, D. (2015). Enhanced twitter sentiment classification using contextual information. *MIT Press* (pp. 1–10).
Wang, Z., & Zhang, Y. (2017). Opinion recommendation using a neural model. In *Proceedings of the Conference on Empirical Methods on Natural Language Processing.*
Wei, F. (n.d.). Sentiment analysis and opinion mining.
Whelan, G., Moon, J., & Grant, B. (2013). Corporations and citizenship arenas in the age of social media. *Journal of Business Ethics, 118*(4), 777–790.
Wiebe, J., Wilson, T., & Cardie, C. (2005). Annotating expressions of opinions and emotions in language. *Language Resources and Evaluation, 39*(2–3), 165–210.
Zhang, L., Wang, S., & Liu, B. (2018). Deep learning for sentiment analysis: a survey.

Israel Edem Agbehadji graduated from the Catholic University College of Ghana with B.Sc Computer Science in 2007, M.Sc. Industrial Mathematics from the Kwame Nkrumah University of Science and Technology in 2011 and Ph.D. Information Technology from Durban University of Technology (DUT), South Africa, in 2019. He is Member of ICT Society of DUT Research Group in the Faculty of Accounting and Informatics. He lectured undergraduate courses in both DUT, South Africa, and a private university, Ghana. Also, he supervised several undergraduate research projects. Prior to his academic career, he took up various managerial positions as the management information systems manager for National Health Insurance Scheme and the post-graduate degree programme manager in a private university in Ghana. Currently, he is Postdoctoral Research Fellow at DUT, South Africa, and working on joint collaboration research project between South Africa and South Korea. His research interests include big data analytics, Internet of things (IoT), fog computing and optimisation algorithms.

Abosede Ijabadeniyi currently works as Postdoctoral Fellow at the Environment Learning Research Centre, Rhodes University, South Africa. She obtained her Ph.D. in Marketing at the Department of Marketing and Retail Management, Durban University of Technology, South Africa, where she also lectured both undergraduate and postgraduate courses. With interdisciplinary research interests which intersect the fields of economics and corporate marketing, she has a keen interest in fostering value proposition for sustainable development based on research into corporate social responsibility (CSR) identity construction and communication. She has publications in accredited journals and has presented papers at local and international conferences and won the City University of New York's Acorn Award for best presentation at the Corporate Communications International Conference in June 2017.

Chapter 10
Data Visualization Techniques and Algorithms

Israel Edem Agbehadji and Hongji Yang

1 Introduction

Visualization is the method of using graphical representations to display information (Ward et al. 2010) in order to assist understanding. Data visualization can be seen as systematically representing data with its data attributes and variables forming the unit of information. Text containing numeric values can be systematically represented visually using traditional tools such as scatter diagrams, bar charts, and maps (Wang et al. 2015). The main goal of a visualization system is to transform numerical data of one type into a graphical representation such that a user becomes perceptually aware of any structures of interest within this data (Keim et al. 1994). Through the depiction of data into the correct type of graphical array (Keim et al. 1994), users are able to detect patterns within datasets. These traditional methods could be challenged, however, with respect the amount of computational time that is needed to visualize the data.

The significance of a bio-inspired behavior, such as dung beetle behavior, for big data visualization is the capability to implement path integration and to navigate with the least amount of computational power. The behavior of a dung beetle, when represented as an algorithm, can find the most appropriate method to visualize discrete data that emerges from various data sources and that needs to be visualized fast with minimal computational time. When less computational time is needed to visual patterns, these patterns can be featured as quickly moving (in conjunction with the

I. E. Agbehadji (✉)
ICT and Society Research Group, Durban University of Technology, Durban, South Africa
e-mail: israeldel2006@gmail.com

H. Yang
Department of Informatics, University of Leicester, Leicester, England, UK
e-mail: Hongji.Yang@Leicester.ac.uk

S. J. Fong and R. C. Millham (eds.), *Bio-inspired Algorithms for Data Streaming and Visualization, Big Data Management, and Fog Computing*, Springer Tracts in Nature-Inspired Computing,
https://doi.org/10.1007/978-981-15-6695-0_10

velocity features of a big data framework). Consequently, with less computational time needed, large amounts of data can be observed using visual formats in the form of graph for easy understanding (Ward et al. 2010).

2 Introduction to Data Visualization

2.1 Conventional Techniques for Data Visualization

Traditional methods for data visualization consider response time and performance scalability during the visual analytics phase (Wang et al. 2015). Response time correlates to the pace (in other words, the velocity features of the big data framework) at which data points appear and how often it alters when there is huge amount of data (Choy et al. 2011). Among the methods of visualization are stacked display method and the dense pixel display method (Keim 2000; Keim 2002; Leung et al. 2016).

Keim (2000) indicated that the idea of dense pixel technique is to map each dimension value, whether numeric or text data, to a colored pixel and then bring together the pixels associated to each dimension into nearby areas using the circle segments method (which is gathering of all features, in proximity to a center and in proximity to one another, to improve the visual comparison of values).

The stacked display methods (Keim 2002; Leung et al. 2016) depicts sequential actions using a hierarchal manner. This hierarchal manner assembles a stack of displays to represent a visual format. The main goal of the stack display is to incorporate one coordinate system inside another such that two attributes constitute the outer coordinate system and two other attributes are then incorporated into the outer coordinate system and so on as each set of attributes are incorporated into their nearest outer layer so that many layers now compose one large layer.

2.2 Big Data Visualization Techniques

In this subsection, we look at big data visualization techniques to handle large volumes, different varieties, and varying velocities of data. In this context, the volume denotes to the quantity of data, variety denotes whether the data is structured, semi-structured or unstructured data, and velocity denotes the speed both needed to receive and analyze data along with how often the data is frequently changed. One challenge with big data is when a user is overwhelmed with results that are not meaningful. After looking at the representation of data vis-à-vis the characteristics of volume, variety, and velocity, we quickly look at their graph database to show to represent relationships among data and a business example as to how these relationships can provide business insights. The following sections present the fundamental techniques to visualize big data characterized by volume, variety, and velocity:

Binning

This technique groups data together in both *x*- and *y*-axes for effective visualization. In the process of binning, billion rows of dataset are grouped into two axes within the shortest possible time. Binning is one of the techniques to visualize volume of data in big data environment (SAS Institute Inc 2013).

Box Plots

Box plots are the use of statistics to summarize distribution of data and present the results using boxes. There are five statistical techniques used in box plot, and these are the "minimum, maximum, lower quartile, median and upper quartile." The box plot technique to visualize data helps to detect outliers in a large amount of data. These data outliers in the form of extreme values are represented by whiskers that extend out from the edges of the box (SAS 2017). The box plot is one of the techniques to visualize volume of data in big data environment.

Treemap

Treemap is a technique to view data in hierarchical manner (Khan and Khan 2011) where rectangles are used to represent data attributes. Each rectangle has unique color with other sub-rectangles to show the measure of data as a collection of choices for streaming music and video tracks in a social network community. The Treemap is one of the techniques to visualize large volumes of data in the big data environment.

Word Cloud

Word cloud uses the frequency of a word in visualizing data, where the size of each word represents the number of occurrences of a word in a text. Word cloud visualization is a technique used to visualize unstructured data and present the results using the high or low frequency of each word. Word cloud visualization is based on the concept of taxonomy and ontology of words in order to create an association between words (SAS 2017). The association between words enables users to drill down further for more information on the word. This approach has been used in text analysis. The word cloud is one of the techniques to visualize the variety of data in big data environment.

Correlation Matrices

Correlation matrices are a technique that uses matrices to visualize big data. The matrix combines related variables and show how strongly correlated one variable is with that of the other. In the process of creating a visual display, color-coded boxes are used to represent data points on the matrices. Each color-codes on a grid/box shows whether there is a strong or weak correlation between variables. Strong correlation may be represented with darker color-codes in boxes while weaker correlation may be indicated with light color-code boxes. The advantage of using correlation matrices is that it combines big data and fast response time to create a quick view of related variables (SAS 2017). The correlation matrices are one of the techniques to visualize varying velocity of data in big data environment.

Parallel Coordinates

Parallel coordinates technique is a visualization technique for high-dimensional geometry is built on projective geometry (Inselberg 1981, 1985; Wegman 1990) where the visualized geometry represents data in multiple domains or attributes (Heinrich 2013). This technique places attributes on axes in parallel with each other such that more dimensions of attributes can be viewed in single plot (Heinrich 2013). Thus, single data elements can be plotted across several dimensions connected to the *y*-axis and each object of the data is shown along the axes as a series of connected data points (Gemignani 2010). Parallel visualization technique can be applied in air traffic control, computational geometry, robotics, and data mining (Inselberg and Dimsdale 1990). The parallel visualization technique is one of the techniques that can be used to visualize data characterized as having volume, velocity, and variety of data in big data environment. The parallel coordinate for five attributes is represented by vertical axis while data point of each attribute is mapped to a polygonal line that intersect each axis with their corresponding coordinate value. The challenge with this technique is the difficulty in identifying data characteristics as many points that are represented on parallel coordinates (Keim and Kriegel 1996).

Network Diagrams

Network diagram is the use of nodes (that is individual actors with the network) and connections (that is the relationships) between each of the nodes (SAS 2017). The network diagram technique, designed to visualize unstructured and semi-structured data, uses nodes to represent data points and connection between each data point as lines. This form of data representation creates a map of data which could help identify interaction within several nodes. For example, the network diagram can be applied for counterintelligence, law enforcement, crime related activities, etc. The network diagram is one of the techniques to visualize volume and variety of data in big data environment.

Graph Databases

Based on network diagrams, many specialized big data databases are being used. Although the traditional relational database is well known for its solid mathematical basis, maturity, and standardization which enables it to remain commonplace for small to medium sized datasets (SyonCloud 2013), these relational databases often cannot handle large datasets (Agrawal et al. 2010). Consequently, many specialized solution databases for large datasets are developed or being developed. In the case of large datasets and of a situation where the relationships between data items (nodes of information) are more significant than the data items themselves specialized graph databases have been developed as a solution. Using key-value properties, both nodes and relationships are referenced and accessed. In order to utilize graph databases, the nodes must be discovered first and then the relationships between the nodes is identified (Burtica et al. 2012).

Business Examples of Using Graphs for Business Insights

Based on network theory, nodes (which indicate individual actors within a network or data items) and relationships (which indicate associations between nodes) are integrated into the graph database. These relationships can indicate employee relationships, email correspondence between node actors, Twitter responses, and Facebook friends. If social networks are represented using a graph database, the nodes and their relationships can be used to define and analyze multiple authorships, client relationships, and corporate structures (Lieberman 2014).

A concrete example of a social network, which is utilized in a real-life business setting, is client purchases in a supermarket. Various food items may be denoted as entities or groups of entities; those items which are purchased together are denoted as relationships with value weighing and transaction rates. The relationship with the heaviest weighing is not always of the most interest to the business manager. An example, it is common knowledge that frankfurters and buns are bought together. The most-valued information may be what is currently unknown such as which entity is common to all transactions. In this case, the most common item may be bread due to its common usage and short shelf life. With this knowledge, a supermarket may decide to attract all types of shoppers by promoting and discounting its bread (Lieberman 2014).

3 Bio-Inspired Technique to Data Visualization

3.1 Cellular Ant Based on Ant Colony System

Moere et al. (2006) combined features of ant and cellular automata to create visual groups on datasets. The combined characteristics are referred as cellular ant. Generally, the cellular ants can create a self-organization structure which helps to independently detect pattern similarity in multi-dimensional datasets. The self-organizational structure dynamically creates visual cues that show position, color, and shape-size of visual objects (that is data points). Due to its dynamic behavior, a cellular ant decides its visual cues independently, as it can adjust to specific color, swap its position with a neighbor, and move around or stay put. In this case, the positional swap correlates to a swapping between data values that are plotted on a grid for the purpose of user visualization. In cellular ant, the structure of individual ants corresponds to the data point. These cellular ants perform a continuous pair-wise negotiation with neighboring ants which then create visual patterns on a data grid. Commonly, there is no direct predefined mapping rule that interconnects visual cues with data values (Moore et al. 2006). Therefore, the shaped-size-scale adjustments automatically adapt to the data scale through self-organization and an autonomous approach. Therefore, rather than map a specific space-size to a data value, each ant in the ant colony system maps one of its data attributes onto its size through negotiation with its neighbors. Through this shaped-size negotiation procedure, ants compare at random their similar data value and each circular radius size, representing a pixel. The self-organizing behavior and negotiation between ants are guided by simplified rules that help to either grow the population of ants by attracting similar ants together or by shrinking the ant's population. These rules are important in defining the scale of visual data whereas the randomized process is important in defining the adaptability of the data value. The procedure of scale and share-size negotiation; however, may entail the need for considerable computational time in order to coordinate the clustering of ants or perform a solitary finished action.

3.2 Flocking Boids System

The flocking boid system is based on the behavior of birds. The swarming movement is steered by simplified mathematical rule that depicts the flocking simulation of birds objects (called boids). Accordingly, boids tend to move as close to the center of the herd as possible, and thus, in visualization terminology, cluster. Such boids act as agents: they are positioned, seeing the world from their own perspective rather than from a global one, and their actions are determined by both internal states as well as external influences (Moere 2004). The rule based behavior of each boid obeys five behavior rules namely Collision Avoidance, Velocity Matching, Flock Centering, Data Similarity and Formation Forming (Moere and Lau 2007).

The rule-based behavior systems can frequently update and continuously control their dynamic actions of individual and create three-dimensional elements that represent the changing data values of reoccurring data objects. The flocking boid system is driven by local interactions between the spatial elements as well as the evolution of time-varying data values. The flocking approach/algorithm that includes time-varying datasets enables continuous data streaming, live database querying, real-time data similarity evaluation, and dynamic shape formulation. Alternative methods of visualizing time-varying datasets include Static State Replacement, Equilibrium Attainment, Control Applications, Time-Series Plots, Static State Morphing, and Motion-Based Data Visualization (Moere 2004). The static state replacement method requires a continuous sequence which can be ineffectually recognized as discrete steps. The static State morphing method requires pre-computation of the static states and is incapable of visualizing real-time data. The Equilibrium Attainment method also requires pre-computation of data similarity matrices and does not create recognizable behavior as the motion characteristics denote no particular meaning. The Control Applications method is quite effective as data streams are aggregated online and gradually streams representative data objects to the visualization system. The Time-Series Plots method employs time series plotting that connects sets of static states and maps these states in space and time with simple drawn curves. The three-dimensional temporal data scatter plots are useful in solving the time-varying data evaluation and visualization performance challenges because of the distributed computing and shared memory parallelism nature used within this approach (Moere 2004).

3.3 Dung Beetle-Based System

The dung beetle possesses a very small brain (comparable in size to a grain of rice). Dung beetle forages on the dung of herbivorous animals. One known characteristic of the dung beetle is their ability to use minimal computational power for orientation and navigation using the celestial polarization pattern (Wits University 2013). Dung beetles can be categorized into three groups: dwellers, tunnelers, and rollers. Dwellers remain on the top of a dung pile to lay their eggs. Tunnelers alight on a heap of dung and burrow down into the heap. Rollers shape the dung into a ball and then roll this newly formed ball to a safe location. The author of Kuhn and Woolley (2013) indicates that a directing principle for visualization is the utilization of simple rules to create multifaceted phenomena. These simple rules pertain to basic rules which govern a dung beetle's dynamic behavior namely: Ball rolling on a straight line; dance based on a combination of internal cue of direction and distance with external reference obtained from its environment and then positioning themselves using the celestial polarized pattern; and the path integration that is sum sequential modification in location in hierarchical fashion and continuously updating of distance and direction from the initial point to return home (Agbehadji et al. 2018).

4 Data Visualization Evaluation Techniques

Although data visualization techniques and their effectiveness are difficult to evaluate objectively, Keim provides the quantitative measuring approach. The approach is based on synthesized test data attributes with similar features, such as data type—integer, float, and string; comparable to that of real dataset where the data value relates to each other by the variance and mean, the size, shape, and position of clusters, and their distribution and the correlation coefficient of two dimensions. Some common features of the data types include metric—data that has an important distance metric between any two values nominal—data whose values have no intrinsic ordering; ordinal—data whose values are ordered, but lacks any no important distance metric. When some parameters (like statistical parameters) that express the data, characteristics are varied with time within a controlled experiment, these varying parameters assist in assessing various visualization methods. In this regard, this experiment determines where the data features are noticed for the first time and when these features are no longer noticed. Consequently, it gives a more realistic test data with diverse parameters on data. Another technique proposed by Keim is to use the same test data when comparing various visualization methods in order to identify the advantages and shortcomings of each method. The use of experiment and "same test data" techniques are subjective because it is based on users' experience and the use of a particular visualization technique.

Another method of finding the effectiveness of visualization technique is based on how it enables the user to read, understand, and interpret the display easily, accurately, quickly, etc. Card et al. 1999, defines efficacy as the ability of human to properly view a display and understand the results more rapidly, and conveys the distinctions in the display with fewer errors. Thus, efficacy is assessed with regards to the quality of the tasks, solution, or to the time required to finish a given task (Dull and Tegarden 1999; Risden and Czerwinski 2000).

Some other visualization evaluation techniques include user observation, implementation of questionnaires, and the use of graphic designers to critique visualized results (Santos 2008) and to give their opinion on them. Though these visualization evaluation techniques are important, it is qualitative and subjective. Consequently, the use of a quantitative approach could supply a more objective means to assess visualization evaluation methods.

5 Conclusion

This chapter reviewed current methods and techniques used in data visualization. Although the computational cost of creating visual data is one of the challenges, the scale of data to be visualized requires the use of other methods. One such solution meant to address the issue of computational cost and scalability of data is the use of nature-inspired behavior. The advantage of nature-inspired methods is the ability to avoid searching through non-promising results to find the most or near optimal search. The simplified rules that can be formulated from their behavior make it easy to understand and implement in any visualization problem. Among such nature-inspired behavior which has been proposed is dung beetle for data visualization. These simplified rules are in the form of mathematical expressions; hence, it can provide an objective way of measuring effectiveness of data visualization technique, which is an area that requires further research.

Key Terminology and Definitions

Data visualization—is the process of representation data in a systematic form with data attributes or variables that represent unit of information.

Data visualization technique—is an approach that transforms data into a format that a user can view, read, understand, and interpret the display results easily, accurately, and quickly.

Bio-inspired/Nature-inspired—refers to an approach that mimics the social behavior of birds/animals. Bio-inspired search algorithms may be characterized by randomization, efficient local searches, and the discovering of the global best possible solution.

References

Agbehadji, I. E., Millham, R., Fong, S. J., & Yang, H. (2018). Kestrel-based search algorithm for association rule mining of frequently changed items with numeric and time dimension (under consideration).

Agbehadji, I. E., Millham, R., Thakur, S., Yang, H. & Addo, H. (2018). Visualization of frequently changed patterns based on the behaviour of dung beetles. In *International Conference on Soft Computing in Data Science* (pp. 230–245).

Agrawal, D., Das, S., & El Abbadi, A. (2010). Big data and cloud computing: New wine or just new bottles? *Proceedings of the VLDB Endowment, 3*(1–2), 1647–1648.

Burtica, R., et al. (2012). Practical application and evaluation of no-SQL databases in cloud computing. In *2012 IEEE International Systems Conference (SysCon)*. IEEE.

Card, S. K., Mackinlay, J. D., & Shneiderman, B. (1999). *Readings in information visualization—Using vision to think*. San Francisco, CA: Morgan Kaufmann Publishers.

Choy, J., Chawla, V., & Whitman, L. (2011). *Data visualization techniques: From basics to big data with SAS visual analytics*. https://www.slideshare.net/AllAnalytics/data-visualization-techniques.

Dull, R. B., & Tegarden, D. P. (1999). A comparison of three visual representations of complex multidimensional accounting information. *Journal of Information Systems, 13*(2), 117.

Etienne, A. S., & Jeffery, K. J. (2004). Path integration in mammals. *Hippocampus, 14,* 180–192.

Etienne, A. S., Maurer, R., & Saucy, F. (1988). Limitations in the assessment of path dependent information. *Behavior, 106,* 81–111.

Gemignani, Z. (2010). *Better know a visualization: Parallel coordinates*. www.juiceanalytics.com/writing/parallel-coordinates.

Golani, I., Benjamini, Y., & Eilam, D. (1993). Stopping behavior: Constraints on exploration in rats (Rattus norvegicus). *Behavioural Brain Research, 53,* 21–33.

Heinrich, J. (2013). Visualization techniques for parallel coordinates.

Inselberg, A. (1981). *N-dimensional graphics* (Technical Report G320-2711). IBM. Cited on page 7.

Inselberg, A. (1985). The plane with parallel coordinates. *The Visual Computer, 1*(4), 69–91. Cited on pages 7,8,18, 25, and 38.

Inselberg, A., & Dimsdale, B. (1990). *Parallel coordinates: A tool for visualizing multi-dimensional geometry* (pp. 361–370). San Francisco, CA: Visualization 90.

Keim, D. (2000). Designing pixel-oriented visualization techniques: Theory and applications. *IEEE Trans Visualization and Computer Graphics, 6*(1), 59–78.

Keim, D. A. (2001). Visual exploration of large data sets. *Communications of the ACM, 44,* 38–44.

Keim, D. A. (2002). Information visualization and visual data mining. *IEEE Transactions on Visualization and Computer Graphics, 8*(1).

Keim, D. A., Kriegel, H. (1996). Visualization techniques for mining large databases: A comparison. *IEEE Transactions on Knowledge and Data Engineering, Special Issue on Data Mining, 8*(6), 923–938.

Keim, D. A., Bergeron, R. D., & Pickett, R. M. (1994). *Test data sets for evaluating data visualization techniques*. https://pdfs.semanticscholar.org/7959/fd04a4f0717426ce8a6512596a0de1b99d18.pdf.

Khan, M., & Khan, S. S. (2011). Data and information visualization methods and interactive mechanisms: A survey. *International Journal of Computer Applications, 34*(1), 1–14.

Kuhn, T., & Woolley, O. (2013). *Modeling and simulating social systems with MATLAB; Lecture 4—Cellular automata*. ETH Zürich.

Leung, C. K., Kononov, V. V., Pazdor, A. G. M., Jiang, F. (2016). PyramidViz: Visual analytics and big data visualization of frequent patterns. In *IEEE 14th International Conference on Dependable, Autonomic and Secure Computing, 14th International Conference on Pervasive Intelligence and Computing, 2nd International Conference on Big Data Intelligence and Computing and Cyber Science and Technology Congress.*

Lieberman, M. (2014). Visualizing big data: Social network analysis. In *Digital Research Conference.*
Lu, C. -T., Sripada, L. N., Shekhar, S., & Liu, R. (2005). Transportation data visualisation and mining for emergency management. *International Journal of Critical Infrastructures, 1*(2/3), 170–194.
Mamduh, S. M., Kamarudin, K., Shakaff, A. Y. M., Zakaria, A., & Abdullah, A.H. (2014). Comparison of Braitenberg vehicles with bio-inspired algorithms for odor tracking in laminar flow. *NSI Journals Australian Journal of Basic and Applied Sciences, 8*(4), 6–15.
Marghescu, D. (2008). *Evaluating multidimensional visualization techniques in data mining tasks.* http://www.doria.fi/bitstream/handle/10024/69974/MarghescuDorina.pdf?sequence=3&isAllowed=y.
Mittelstaedt, H., & Mittelstaedt, M.-L. (1982). Homing by path integration. In F. Papi & H. G. Wallraff (Eds.), *Avian navigation* (pp. 290–297). New York: Springer.
Moere, A. V. (2004). Time-varying data visualization using information flocking boids. In *IEEE Symposium on Information Visualization* (p. 8).
Moere, A. V., & Lau, A. (2007). Information flocking: An approach to data visualization using multi-agent formation behavior. In *Proceedings of Australian Conference on Artificial Life* (pp. 292–304). Springer.
Moere, A. V., Clayden, J. J., & Dong, A. (2006). *Data clustering and visualization using cellular automata ants.* Berlin Heidelberg: Springer.
Risden, K., & Czerwinski, M. P. (2000). An initial examination of ease of use for 2D and 3D information visualizations of web content. *International Journal of Human—Computer Studies, 53,* 695–714.
Santos, B. S. (2008). Evaluating visualization techniques and tools: What are the main issues? http://www.dis.uniroma1.it/beliv08/pospap/santos.pdf.
SAS Institute Inc. (2013). *Five big data challenges and how to overcome them with visual analytics.* Available http://4instance.mobi/16thCongress/five-big-data-challenges-106263.pdf.
SAS Institute Inc. (2017). *Data visualization techniques: From basics to big data with SAS visual analytics.* sas.com/visual-analytics.
Synocloud. (2013). *Overview of big data and NoSQL technologies as of January 2013.* Available at http://www.syoncloud.com/big_data_technology_overview. Accessed 22 Dec 2015.
Wang, L., Wang, G., & Alexander, C. A. (2015). Big data and visualization: Methods, challenges and technology progress. *Digital Technologies, 1*(1), 33–38. Science and Education Publishing Available online at http://pubs.sciepub.com/dt/1/1/7.
Ward, M., Grinstein, G., & Keim, D. (2010). *Interactive data visualization: Foundations, techniques, and application,* A K Peters.
Wegman, E. J. (1990). Hyper dimensional data analysis using parallel coordinates. *Journal of the American Statistical Association, 85*(411), 664–675. Cited on pages 7, 8, 9, 18, 38, 39, and 101.
Wits University. (2013). *Dung beetles follow the milky way: Insects found to use stars for orientation.* ScienceDaily. https://www.sciencedaily.com/releases/2013/01/130124123203.htm.

Israel Edem Agbehadji graduated from the Catholic University college of Ghana with B.Sc. Computer Science in 2007, M.Sc. Industrial Mathematics from the Kwame Nkrumah University of Science and Technology in 2011 and Ph.D. Information Technology from Durban University of Technology (DUT), South Africa, in 2019. He is a member of ICT Society of DUT Research group in the Faculty of Accounting and Informatics. He lectured undergraduate courses in both DUT, South Africa, and a private university, Ghana. Also, he supervised several undergraduate research projects. Prior to his academic career, he took up various managerial positions as the management information systems manager for National Health Insurance Scheme; the postgraduate degree programme manager in a private university in Ghana. His research interests include big data analytics, Internet of things (IoT), fog computing, and optimization algorithms. Currently, he works as a Postdoctoral Research Fellow, DUT-South Africa, on joint collaboration research project between South Africa and South Korea.

Hongji Yang graduated with a Ph.D. in Software Engineering from Durham University, England, with his M.Sc. and B.Sc. in Computer Science completed at Jilin University in China. With over 400 publications, he is full professor at the University of Leicester in England. Prof Yang has been an IEEE Computer Society Golden Core Member since 2010, an EPSRC peer review college member since 2003, and Editor in Chief of the International Journal of Creative Computing.

Chapter 11
Business Intelligence

Richard Millham, Israel Edem Agbehadji, and Emmanuel Freeman

1 Introduction

In this chapter, we first look at patterns with their relevance of discovery to business. We then do a survey and evaluation, in terms of advantages and disadvantages, of different mining algorithms that are suited for both traditional and big data sources. These algorithms include those designed for both sequential and closed sequential pattern mining, as described in previous chapters, for both the sequential and parallel processing environments.

2 Data Modelling

The modern relational model arose as a result of issues with existing hierarchical-based indices, data inconsistency, and the need for data independence (separate applications from their data to permit data growth and mitigate the effects of changes in data representation). The variety of indexing systems used by various systems often become a liability as it requires a corresponding number of similar applications that

R. Millham (✉) · I. E. Agbehadji
Society of ICT Group, Durban University of Technology, Durban, South Africa
e-mail: richardm1@dut.ac.za

I. E. Agbehadji
e-mail: israeldel2006@gmail.com

E. Freeman
Faculty of Computing and Information Systems, Centre for Online Learning and Teaching (COLT), Ghana Technology University College, Accra, Ghana
e-mail: efreeman@gtuc.edu.gh

S. J. Fong and R. C. Millham (eds.), *Bio-inspired Algorithms for Data Streaming and Visualization, Big Data Management, and Fog Computing*, Springer Tracts in Nature-Inspired Computing,
https://doi.org/10.1007/978-981-15-6695-0_11

are able to manage these indices and index structures (Codd 1970) The relational model keeps related data in a domain within a table with links to sub-domains of data in other tables and links to other domains of data (which are also represented as tables) In addition, with traditional file systems come to the possibility of inconsistency as a given domain of data, such as addresses, may be duplicated in several files yet updates may only affect a subset of these files. Codd's relational model has mechanisms to detect redundancy as well as several proposed mechanisms to manage identified redundancies (Codd 1970). Although the relational model separates data from application in order to allow data independence, it comes at a cost of multiple linked tables with little real meaning to end-users.

Most modern data modelling tools are based on structural notations. These notations concentrate on interconnected information entities that detail a domain data structure. Normalization, a logical view mechanism, reduces the understanding of the model by an end-user. As well, systematic documentation of this model at all abstraction levels is needed (Kucherov et al. 2016). Some issues with data modelling using structural notation are denoted as the following:

An end-user does not view data as a separate entity distinct from the methods of their "creation, transformation and processing" in many instances. The end-user forms the foundation of their understanding of data on subject orientation (Rogozov 2013). Subject orientation, rather than concentrating on the data item itself as a distinct object, emphasizes on users' activities which produce, transform and process this data and view information itself as the end result of a user process.

Normalization, with its focus on linked tables with no redundant information, make users' perception of data difficult in that the user must understand a multitude of table with their linkages in addition to the domain data object semantic description.

There is a lack of a causal link between processing (the production of) data and their result. The processes, which may produce or transform data, are connected with the business logic of the organization and the data it handles (Kucherov et al. 2016).

In order to mitigate these limitations, an end-user subject-oriented data modelling is proposed. Data is an integral part of processes which cannot be detached. A data item is categorized by a group of attributes that define values of its chief properties. A concept of a data object itself is categorized by a description of its process which, when implemented, form its understanding. To differentiate between a concept and a data item/object, an example of a book is used:

A data object of a book is denoted by its group of attributes: title, author, number of pages, kind of binding, et al.

A concept of a book will include a description of the processes that produce a book form its cover to its pages of text (which include attributes such as title and author). The pages of text themselves are concepts which include the process of their production of the text itself and rules for its printing.

This subject orientation of data item has three levels, based on their context. If the engaged process of a data item has no particular values, this concept becomes implicit. It may be known that a data item is a book but we do not know its format, in terms of it being paper or electronic, or whether or not it is bound, et al. A concept sense contains concrete characteristics during the description of a process such as its page size, cover type. During this process implementation, a "concept with explicit sense" is acquired. This concept relates to a data object but has only a description of its expected result (in this instance, a book) and the process of its production. If a defined process is implemented, its features will contain particular values and the concept itself becomes definite (Kucherov et al. 2016).

In order to distinguish between subject- and object-oriented approaches using these levels, it is important to remember that the object-oriented concepts have no implicit sense, like the subject approach, but have concrete concepts and concepts with an explicit sense. An example, in the object-oriented approach, an object, BankAccount, will represent an instance of account in a bank, a concrete concept. Furthermore, this BankAccount object will have methods and attributes associated explicitly with a bank account, such as a BankBalance attribute or Withdraw method.

Subject-oriented concepts will have elements, tools, results and functions dependent on the concept's sense. With a concept with an explicit sense, an action is a reflection of this concept and can be denoted by elements, functions, tools (that control the rules for the implementation of functions on elements) and results (through implementing an action according to the purposes of its implementation). An action may indicate a result as per expected target with the actual result showing up after action implementation (Kucherov et al. 2016).

An action may be represented as the basic unit of data storage. An example of a user's action is "payment of monthly wage to worker". In terms of elements, it uses information on the number of days worked (which in turn is calculated by "counting" the days worked and with information about the cost of one working day). In terms of functions, it uses the "mathematical operator multiplication". The tool uses the multiplication rule that is enclosed in one of the system modules that are utilized. Other parts of data storage include functions and results (Kucherov et al. 2016).

The role of a subject-oriented data model is using a single depiction for stored elements and their links amongst them. The end-user is able to read, from the database, the semantics of a data item with the nature of their happening. This new perspective has a beneficial effect both on the modelling process and on the operation and improvement of the system by the user. A subject-oriented approach eliminates the "separation of data, logic and interface" while permitting the notation to illustrate clearly a structure of user data and the results of users manipulating this data within a system (Kucherov et al. 2016).

While a subject-oriented notational approach may show the user how a data element is manipulated (and thus, provide a long-sided semantic meaning to the data), it is rather complex for even small databases, assumes user has knowledge of all data manipulation processes illustrated, and does not clearly delineate between separate data using the same processes. An example, it is not clear how monthly salary (which, in many countries, is a misnomer as it is not dependent on days worked—such an example, might be better suited to a daily rate worker) could be distinguished from one worker to another.

3 Role of Data Mining in Providing Business Insights in Various Domains

Big data may be utilized in many different ways to gain business insights and advantage. Many businesses use IT and big data, for continuous business experimentation that leads test business models, products and improvements in client experience. Big data may assist organizations to make decisions within real time (Bughin et al. 2010).

Lavalle argues that relying on using the whole set of data to derive insights is problematic as this process often takes too much time such that by the time that the first insight is delivered, it is too late. Instead, companies should focus on a specific subject area and the required insights, to be obtained from big data, to meet a specific business objective (Lavalle et al. 2011).

In terms of big data, organizations would commonly begin with well-defined goals, look at well-specified growth objectives, and then design approaches to the more difficult business challenges. In an aspirational company, approximately one-half of business analytics was used by financial management with one third being used for operations, sales and marketing purposes. This approach is common for traditional method of adopting data analytics in inherently data-intensive areas of a business. Experienced companies used big data analytics for the same purposes as aspirational company but at a larger level. Two-thirds of data analytics approach was used by finance with other operational areas such as strategy, product research and customer service development. At a transformed company, business analytics was used in the same areas as an experienced company but it was also used for more difficult areas, such as customer service, to retain and cultivate new customers. Furthermore, success using business analytics often inspired adoption in areas. An example, if business analytics improved supply-chain management, human resources were more likely to use it for workforce planning (Lavalle et al. 2011).

One method analyses financial news articles, collected over a period of one month, which focus on the stock market. This analysis uses the tone and sentiment of words in these articles, along with machine intelligence for interpretation, to develop a financial system to accurately forecast future stock price (Schumaker et al. 2012).

Many corporations gather huge amounts of transactional data and correlate them to customers via their loyalty card reward programme in order to analyse this data for new business opportunities. Such opportunities are developing the most useful promotions for a given customer segment or to obtain critical insights that guide decisions on pricing, shelf distributions and promotions. Often this is done on a weekly basis but an online grocer, Fresh Direct, increases its frequency of decision-making to a daily or more frequent basis, based on data feeds from its online transactions from clients to its web site and on interactions from customer service. Based on this frequency, Fresh Direct adjusts its prices and promotions to its consumers in order to quickly adjust to an ever-changing market (Bughin et al. 2010).

Big data has influence data analysis approach in the energy sector. The influence is because traditional databases are not adapted to process huge volume of both structured and unstructured data. In view of this, the paradigm of big data analytics has become very relevant for the energy sector (Munshi and Yasser 2017). For instance, smart metering systems have been developed to leverage on big data and to enable automated collection of energy consumption data (Munshi and Yasser 2017). The data on consumption enables efficient, reliable and sustainable analysis of a smart grid. However, the massive amounts of "data evolving from smart grid metres used for monitoring and control purposes need to be sufficiently managed to increase the efficiency, reliability and sustainability of the smart grid". Munshi and Yasser (2017) presented a smart grid data analytics "framework on secure cloud-based platform". The framework allows businesses to gain insight into energy demands for a "single-house and a smart grid with 6000 smart metres".

Big data enables a prediction of power outages, system failures and the ability to optimize the utilities, equipment and propose budgets on maintenance. This optimization is achieved through the use of optimization algorithms with their equipment inventory, along with equipment lifecycle history, to optimize resource allocation. Prediction of power outages is provided through analysis of past power outages with their causes in comparison with current circumstances (Jaech et al. 2018). Thus, a utility company leverages its records of its current equipment, equipment maintenance history, and equipment types and their failures and their data analysis for optimization purposes. Similarly, it is able to analyse records of past power outages with their complex causes in order to provide a prediction model (Tu et al. 2017). Big data has been applied to enhance operating efficacy of the power distribution, generation and the transmission; it has been applied to develop a "tailor-made" energy service on a given power grid for different consumers that is both domestic and commercial users; to forecast consequences of the integration of renewable energy sources into the main power grid; and for timely decision-making of top managers, employees, consumers on issues of energy generation and distribution (Schuelke-Leech et al. 2015). Consequently, analysis of big data specific to a utility company can be utilized in a number of ways in order to gain numerous business insights and models that can be used to enhance this particular business' processes.

Big data is often used in conjunction with the Internet of things to determine the circumstances of a situation and make adjustments accordingly. An example, many insurers in Europe and the USA install sensors in client vehicles to monitor driving patterns. Based on the information gained from these sensors (after being processed by big data methods), it allows insurers to give new pricing models that use risk based on driving behaviour rather than on a driver's demographic features. Another example occurs often in manufacturing where sensors continually take detailed readings of conditions at various stages of the manufacturing processes (whether the manufacture concerns computer chips or pulp and paper) and automatically makes modifications to mitigate downtime, waste and human involvements (Bughin et al. 2010).

Businesses often use big data to produce value via incremental and radical innovation (Story et al. 2011). An example, Google might use big data, which correlates an advert displayed on a smartphone of a user during an internet search and geolocation of the phone, actually resulted in a store visit (Baker and Potts 2013) Such correlations, labelled as insights, are frequently used to assess and improve the efficacy of digital advertising (Story et al. 2011). Improving the effectiveness of advertising and obtaining a better understanding of customers may lead to incremental innovation of a business (Story et al. 2011). However, this is often insufficient as incremental innovation, though needed, is not enough to attain a sustainable competitive advantage over rivals (Porter 2008). The customer insights, which are acquired through big data mining, must be used to constantly reshape an organization's marketing and other activities in order to institute radical innovation (Tellis et al. 2009).

Adaptive capability is the ability of organizations to predict market and consumer trends. This capability often is derived from gathering consumer activities and extract undiscovered insights (Ma et al. 2009). Adaptive capability, along with the ability to respond dynamically to change, motivate innovation and allows organizations to develop further value (Liao et al. 2009). An example of adaptive capability, which leads to innovative operational change, is the adoption of anticipatory shipping by Amazon. Amazon mines big data by sifting through a client's order history, "product search history and shopping cart activities" in order to forecast when this client will buy an item online and then, based on this forecast, begin shipping the item to the client's nearest distribution hub before the order is actually place (Ritson 2014). As a result of this forecast, shipping times from Amazon to client are reduced and customer satisfaction increases. These client discernments, obtained from big data, guided Amazon to redevelop its product distribution strategy rather than simply improve them. These types of redevelopment allow firms to use big data to create greater value to their organizations than if they merely adopted an incremental innovation approach (Kunc and Morecroft 2010).

Lavalle categories companies into their categories according to their usage of big data: "aspirational, experienced and transformed". Aspirational companies use big data analytics to justify actions while focusing on cost efficiency with revenue growth being secondary. Experienced companies use these analytics to guide actions but focus on revenue growth with cost efficiency being secondary. Transformed companies use these analytics to recommend actions with primary importance given to revenue growth along with a strong focus on retaining/gaining new clients (Lavalle et al. 2011).

4 Thick Data

Thick data is a multi-disciplinary approach to knowledge that can be obtained from the intersection of "Big" (as in computational) and "Small" (as in ethnographical) data (Blok and Pedersen 2014). In the twentieth century, the study of social and cultural phenomena focused on two types of data. One kind of data was focused on large groups of people which entailed quantitative methods such as statistical, mathematical or computational methods for data analysis. This type of data was ideal for economics or marketing research. The other kind of data focus was focused on a few individuals or small groups and entailed qualitative methods of data analysis. This type of data was commonly used for ethnography and psychology (Manovich 2011). Although big data can quantify human behaviour (as in "how much"), among other things, it cannot explain its motivations (as in "why") (Rassi 2017). Rasmussen argues that big data is very capable of providing answers to well-defined questions or using models based on historical data but this capability is limited to the extent that the modeller selected the accurate types of data to include for analysis and selected the correct assumptions (Rasmussen and Hansen 2015). Cook argues that big data often entails companies becoming too engaged with numbers while neglecting the human requirements of their clients' lives (Cook 2018).

An example of "thick data" supplementing the meaning of discoveries uncovered by big data can be illustrated by the case of a large European supermarket chain that suffered from disappearing market share. The supermarket executives could see the decreasing market share in both the sale figures and that their client's big weekend trips to the market, one of the chief components of their business, seem to be vanishing. However, they were clueless as to what was creating this change. To try to understand these changing phenomena, they tried the traditional marketing approach—a survey of over 6000 shoppers in each market with questions from shopping decisions, price sensitivity, brand importance and motivations to purchase. However, this survey was inconclusive and did not yield any proper insights into the matter. While people indicated that price was an important factor, 80% of respondents prefer high quality over low quality, irrespective of the price. Furthermore, 75% of the respondents mentioned that they shopped at discount stores. These responses created a paradox: if the chain was losing clients to discount stores, why would people state that they would pay for quality? In order to gain a better understanding of this paradox, the chain commissioned a "thick data" study which would produce insights regarding shopping through "spending time with consumers in their homes and daily lives".

Consequently, a team of "thick data", mostly from the social sciences, researchers spent two months with a select group of customers and watched them as they planned, shopped and dined. The results of the study indicated that their not only had their food habits changed but that people's social lives had completely changed. The stability of family routines was gone, most noticeably the vanishing of the traditional family meal on weekdays. Families no longer ate together at the same time and many families had three or four different diets to consider. These social changes had a tremendous effect on shopping behaviour. On average, people shopped more than nine times a week with one person shopping three times per day. Shoppers were not loyal to particular supermarkets but selected the supermarket that was best-suited for their requirement of fast, convenient shopping. After working all day, shoppers did not want to spend time carefully considering different prices at different supermarkets to find the best deal. In terms of quality, the supermarket's assumption of price versus quality proved to be false. These shoppers did not group supermarkets by discount or by premium quality but rather by the mood and their experience of the stores. Some consumers preferred shops that gave the impression of efficiency; others liked fresh and local; and still others choose stores that offered everyday good value. In response, the supermarket management team had to create a shopping experience that was both convenient and unique (Rasmussen 2015).

To confirm the insights gained from this in-depth study, the results were cross-checked against big data from the supermarket's stores. Data on store location and shopping volume for specific stores were correlated in order to provide insight into the significance of convenience. This correlation yielded an insight: the most successful stores were situated in areas where the traffic was the densest, especially in suburban areas. The highest-yielding stores also had a high sense of distinctness designed to fit in with the demographics of their adjacent area. As the supermarket stores were not set up for these new realities, the supermarket's future strategy was focused on an idea in synchronization with what was discovered by this study: developing a distinctive shopping experience that blended well into their customer's fragmented lives (Rasmussen 2015).

These social changes, uncovered by the supermarket management, were also confirmed by Rassi where it was discovered that people would stop in at a grocery store for different reasons—parents who came in to pick up a quick dinner on the way home from soccer practice, people who came in to pick up medicine for an elderly parent, people who tried to get as much groceries from their remaining money before payday, or people who decided to pick up something special to celebrate a big moment in their lives (Rassi 2017).

Another example of a company with decreasing sales due to the lack of engagement with their customers is Lego. Lego, which had enjoyed huge previous successes in their business of producing children's toys, was facing near collapse in the early 2000s. In order to find out why, their CEO, Jorgen Vig Knudstorp, ordered a major qualitative research project which involved studying children in five major global cities in order to better comprehend the emotional needs of children with respect to Legos. While examining hours of video recordings of children at play, a pattern became apparent. They found out that children were fervent about their "play experience" and the process of playing with the consequent activities of imagining and creating. These children did not like the instant gratification of toys like action figures, which Lego had been heavily promoting. Given this feedback, Lego resolved to go back to its traditional building blocks with less attention paid to action figures and toys. As a result, Lego is now a successful company due to its use of thick data (Cook 2018).

Big data may provide information on marketing success, such as the fact that Samsung in 2013 sold 35 million more smartphones than Apple, but this information provides little value. The important question is why Samsung is more popular than Apple? Using thick data, a company can delve into this question. They might find that Apple smartphones lack the range of colours that Samsung provides or are less durable than Samsung. They may find that consumers buy Samsung because it offers a multitude of models that you can customize to your preference with Apple's offerings being less diverse. Through the use of thick data to understand customers' reasons for buying a product is critical for a successful business to maintain its market share or for a failing one to reinvent itself to gain dominance (Cook 2018).

Another example of the limitations of big data and the non-use of "thick data" was the US presidential election of 2016. The traditional polls relied on the accuracy of old models of voting; in doing so, the polls missed some of the significant cultural shifts that occurred that reduced the accuracy of the models upon which the polls were based. The surprise that the "Trump win" generated was because these polls relied on historical voting behaviour of a particular district rather than examining an increasing voter frustration with established institutions which would have been more predictive than this historical voting data. Given this example, the argument that thick data, to help us understand phenomena that are not well-defined, is needed for a fuller picture of reality and to capture insights that traditional big data might miss (Rassi 2017).

5 Challenges

Some challenges with using data analytics for business insights became apparent, when used by human resource departments of companies to recruit applicants, such as gender bias and hiring unqualified staff. Many companies, such as Goldman Sachs and Hilton, are beginning to depend on analytics to help computerize a portion of the recruitment process. One such company was Amazon, whose headcount in 2015 was 575,700, and who was poised to hire more staff (Dastin 2018).

Amazon developed technology that would crawl the web in order to find people whom their analytics deemed worth recruiting. In order to achieve this technology, analytical models were developed that, although particular to a given job function or location, recognized 50,000 terms which appeared on past candidates' CVs from a historical database. The models purposively were designed to allocate little importance on skills that might be pervasive across IT applicants, such as proficiency in a particular programming language. Rather the model placed a great deal of importance on action verbs, such as "executed" or "captured" (Dastin 2018).

A number of challenges immediately emerged. Since this historical database was composed of mostly male applicants (due to hiring practices and cultural norms of the past), words that would be more commonly used by males as opposed to females were predominant and these words were relied on for recruiting decisions. This model clearly demonstrated a "gender" bias against female applicants (Dastin 2018). Furthermore, many of the terms in which the models placed great importance on, such as "executed", were often generic and used in a variety of occupations. Furthermore, while placing little significance on particular programming or technical skills, people who were often totally unqualified for the specific job, were hired. Consequently, after these analytical models produced unsuitable recruits at random, Amazon abandoned the project (Dastin 2018).

Another challenge emerges with respect to privacy of consumers whose buying patterns are uncovered during data mining. A well-known example includes target stores in the USA. Target, while mining its consumers' buying habits and comparing these habits to known patterns, was able to predict that a specific client was pregnant and consequently mailed a flyer promoting their baby products to her home. Although this prediction turned out to be true, this client was still of secondary school age and her family was unaware of her condition until Target's flyer arrived (Allhoff and

Henschke 2018). Allhoff asserts that mining supposedly innocuous data gathered from IOT devices can reveal potentially private information about the owner. An example, the EvaDrop shower (an IOT device that sends data continuously to its manufacturer about its usage) could reveal an unusual increase in shower activity during a specific day. This data, when mined, could indicate that the owner had company over that day. Similarly, if this increase is repeated at regular intervals, such as Saturday morning, it could reveal information that a client may not want others to know (Allhoff and Henschke 2018).

6 Conclusion

In this chapter, we first looked at different data modelling approaches from relational to object-oriented to end-user-oriented modelling. We also looked at how big data could be used within companies for business insights, at different levels, with transformative effects on business processes. Thick data was looked at, with many examples, as a way of complementing the quantitative insights produced by big data for richer insights into client patterns and business processes.

7 Key Terminology and Definitions

Business intelligence—refers to the process of using technology, application software and practices to collect, integrate, analyse and present business information to support decision-making. The intelligence gathered is then presented in business report documents.

Data mining—is the process of finding hidden and complex relationships present in data with the objective to extract comprehensible, useful and non-trivial knowledge from large datasets.

Big data—is a definition that describes huge volume and complicated data sets from various heterogeneous sources. Big data is often known by its characteristics of velocity, volume, value, veracity and variety.

Business Insights—in this chapter, business insights may be referred to as general discernments regarding any facet of the business which is obtained through big data. An example, analysis of transactional and other data (big data) from a supermarket may indicate a trend or pattern (young adults frequent convenience stores after midnight) which may assist the affected business in making a decision to enhance or grow their business. An example of such a decision might be to offer products, or better discounts on these products after midnight, that appeal to this market segment in the store or to offer activities that this segment may be interested in, in order to attract further clients from this segment to the store.

Thick Data—ethnographical or social science data, often obtained via qualitative means, which complement big data's insights and often provide a richer understanding of the insight. An example, analysis of big data might indicate that a certain restaurant chain has lost a specific market segment, 18–28 year olds, but traditional investigative methods, such as surveys, are inconclusive as to the reasons why. A social science approach might be employed to determine the actual reasons for this loss.

References

Allhoff, F., & Henschke, A. (2018). The Internet of Things: Foundational ethical issues. *Internet of Things, 1,* 55–66.

Baker, P., & Potts, A. (2013). 'Why do white people have thin lips?' Google and the perpetuation of stereotypes via auto-complete search forms. *Critical Discourse Studies, 10,* 187–204.

Blok, A., & Pedersen, M. A. (2014). Complementary social science? Quali-quantitative experiments in a Big Data world. *Big Data & Society, 1,* 2053951714543908.

Bughin, J., Chui, M., & Manyika, J. (2010). Clouds, big data, and smart assets: Ten tech-enabled business trends to watch. *McKinsey Quarterly, 56,* 75–86.

Codd, E. F. (1970). A relational model of data for large shared data banks. *Communications of the ACM, 13,* 377–387.

Cook, J. (2018). *The power of thick data.* Big Fish Communications.

Dastin, J. (2018, October 10). Amazon scraps secret AI recruiting tool that showed bias against women. Business News. Available at https://www.reuters.com/article/us-amazon-com-jobs-automation-insight/amazon-scraps-secret-ai-recruiting-tool-that-showed-bias-against-women-idUSKCN1MK08G. Accessed October 10, 2018.

Jaech, A., Zhang, B., Ostendorf, M., & Kirschen, D. S. (2018). Real-time prediction of the duration of distribution system outages. *IEEE Transactions on Power Systems*, 1–9. https://doi.org/10.1109/tpwrs.2018.2860904.

Kucherov, S., Rogozov, Y., & Sviridov, A. (2016). The subject-oriented notation for end-user data modelling. In *2016 IEEE 10th International Conference on Application of Information and Communication Technologies (AICT)* (pp. 1–5). IEEE.

Kunc, M. H., & Morecroft, J. D. (2010). Managerial decision making and firm performance under a resource-based paradigm. *Strategic Management Journal, 31,* 1164–1182.

Lavalle, S., Lesser, E., Shockley, R., Hopkins, M. S., & Kruschwitz, N. (2011). Big data, analytics and the path from insights to value. *MIT Sloan Management Review, 52,* 21.

Liao, J., Kickul, J. R., & Ma, H. (2009). Organizational dynamic capability and innovation: An empirical examination of internet firms. *Journal of Small Business Management, 47,* 263–286.

Ma, X., Yao, X., & Xi, Y. (2009). How do interorganizational and interpersonal networks affect a firm's strategic adaptive capability in a transition economy? *Journal of Business Research, 62,* 1087–1095.

Munshi, A. A., & Yasser, A. R. M. (2017). Big data framework for analytics in smart grids. *Electric Power Systems Research, 151*, 369–380. Available from https://fardapaper.ir/mohavaha/uploads/2017/10/Big-data-framework-for-analytics-in-smart-grids.pdf.

Manovich, L. (2011). Trending: The promises and the challenges of big social data. *Debates in the Digital Humanities, 2,* 460–475.

Porter, M. E. (2008). *On competition.* Boston: Harvard Business Press.

Rasmussen, M. B., & Hansen, A. W. (2015). Big Data is only half the data marketers need. *Harvard Business Review*, 16.

Rassi, A. (2017). *Intended brand personality communication to B2C customers via content marketing.*
Ritson, M. (2014). Amazon has seen the future of predictability. *Marketing Week, 10.*
Rogozov, Y. (2013). Approach to the definition of a meta-system as system. *Proceeding of ISA RAS-2013, 63*, 92–110.
Schumaker, R. P., Zhang, Y., Huang, C.-N., & Chen, H. (2012). Evaluating sentiment in financial news articles. *Decision Support Systems, 53,* 458–464.
Story, V., O'Malley, L., & Hart, S. (2011). Roles, role performance, and radical innovation competences. *Industrial Marketing Management, 40,* 952–966.
Schuelke-Leech, B. A., Barry, B., Muratori, M., & Yurkovich, B. J. (2015). Big Data issues and opportunities for electric utilities. *Renewable and Sustainable Energy Reviews, 52,* 937–947.
Tellis, G. J., Prabhu, J. C., & Chandy, R. K. (2009). Radical innovation across nations: The preeminence of corporate culture. *Journal of Marketing, 73,* 3–23.
Tu, C., He, X., Shuai, Z., & Jiang, F. (2017). Big data issues in smart grid: A review. *Renewable and Sustainable Energy Reviews, 79,* 10991107.

Richard Millham is currently an associate professor at the Durban University of Technology in Durban, South Africa. After thirteen years of industrial experience, he switched to academe and has worked at universities in Ghana, South Sudan, Scotland, and the Bahamas. His research interests include software evolution, aspects of cloud computing with m-interaction, big data, data streaming, fog and edge analytics, and aspects of the Internet of things. He is a Chartered Engineer (UK), a Chartered Engineer Assessor, and Senior Member of IEEE.

Israel Edem Agbehadji graduated from the Catholic University College of Ghana with B.Sc. Computer Science in 2007, M.Sc. Industrial Mathematics from the Kwame Nkrumah University of Science and Technology in 2011 and Ph.D. Information Technology from the Durban University of Technology (DUT), South Africa, in 2019. He is a member of ICT Society of DUT Research group in the Faculty of Accounting and Informatics; and *IEEE* member. He lectured undergraduate courses in both DUT, South Africa, and a private university, Ghana. Also, he supervised several undergraduate research projects. Prior to his academic career, he took up various managerial positions as the management information systems manager for National Health Insurance Scheme; the postgraduate degree programme manager in a private university in Ghana. Currently, he works as a Postdoctoral Research Fellow, DUT, South Africa, on joint collaboration research project between South Africa and South Korea. His research interests include big data analytics, Internet of things (IoT), fog computing and optimization algorithms.

Emmanuel Freeman has M.Sc. in IT, B.Sc. in IT; and PgCert IHEAP from Coventry University, UK. He is a Ph.D. Candidate in Information Systems at the University of South Africa, South Africa. He has seven years teaching and research experience in Information Technology and Computer Science Education. Currently, He is the Head of Centre for Online Learning and Teaching (COLT) and a lecturer at the Ghana Technology University College. His research interest includes information systems, computer science educations, big data, e-learning, blended learning, open and distance learning (ODL), activity-based learning, software engineering, green computing and e-commerce.

Chapter 12
Big Data Tools for Tasks

Richard Millham

1 Introduction

In this chapter, we look at the role of tools in the big data process, particularly but not restricted to the data mining phase.

2 Context of Big Data Being Considered

In order to understand which tool might be most appropriate for a given need, one needs to understand the context of big data, including the users that might utilise the tool, the nature of the data, and the various phases/processes of big data that certain tools might address.

3 Users of Big Data

Many different types of users might consider the use of big data tools. These users include those involved in:

(a) Business applications: these users may be termed the most frequent users of these tools. These tools tend to be various commercial tools that support business application that are linked to databases with large datasets or that are deeply ingrained within a business' workflow.

R. Millham (✉)
Society of ICT Group, Durban University of Technology, Durban, South Africa
e-mail: richardm1@dut.ac.za

S. J. Fong and R. C. Millham (eds.), *Bio-inspired Algorithms for Data Streaming and Visualization, Big Data Management, and Fog Computing*, Springer Tracts in Nature-Inspired Computing,
https://doi.org/10.1007/978-981-15-6695-0_12

(b) Applied research: theses users tend to apply certain tools for data mining or prediction techniques for solutions to specific research problems, such as those found in life sciences. These users may desire tools with well-proven methods, a graphical user interface for quick operation, and various interfaces to link up domain-related data formats or databases.
(c) Algorithm Development: these users will often develop new data mining, or related, algorithms. These users are interested in tools that integrate their newly development methods and evaluate them against existing methods. These tools should contain many concurrent algorithms and libraries of algorithms to aid quick implementation.
(d) Education: for these users, these tools should have a very easy-to-use, interactive interface, be inexpensive, and be very intuitive with the data. These tools should also be integrated with existing learning systems, particularly online, and enable users to be quickly trained on them (Al-Azmi 2013).

4 Different Types of Data

Another aspect to consider when choosing a data mining tool is the dimensionality nature of the data that the tool is processing. Traditionally, data mining tools focused on dealing with two-dimensional sets of data in the form of records in tables. An example, a dataset would contain N instances (such as students within a school) with m characteristics that have real values or symbols (e.g., letter grades for a student's grades). This record-based format is supported by almost all existing tools. Similar dimensionality may occur in different types of datasets. An example would be an n-gram or the frequency of a word within a given text document. Higher dimensional data often have time series as elements with varying dimensions—one instance of a time series with N samples or N various instances of k-dimensional vector time series with K samples. Some examples of these higher dimensional datasets include financial data, energy consumption, and quality inspection reports. Tools that utilise this data typically use this data to forecast future values, group common patterns in a time series, or identify the time series through clustering. This typical use is supported by most data mining tools. Specialised tools are designed to manage various types of structured data such as gene sequences (spatial structuring) or mass spectrograms (which are arranged by masses or frequencies). An emerging trend is data mining among images or videos such as biometric, medical images, camera monitoring, et al. This data, besides having high dimensionality, has the additional problem of huge quantity. Often this data must be split into metadata containing links to image and video files with a specialised tool, such as ImageJ or ITK, processing the images into segmented images and another tool, working in concert, mining these images for patterns (Al-Azmi 2013).

5 Different Types of Tasks

In order to understand where the operation of a given tool fits in the big data processing dataset, it is important to understand what these tasks are.

In terms of grouping together similar items (clustering) and labelling (classification), a number of techniques are used including

(a) Supervised learning—learning done with a known output variable Supervised learning is often utilised for

  a. Classification—labelling of identified classes or clusters.
  b. Fuzzy classification—labelling of data items with their gradual memberships in classes based on their classification values varying from 0 to 1.

(b) Unsupervised learning—learning performed without a known output variable in the dataset. This unsupervised learning often includes

  a. Clustering—identify similarities among data items and groups similar items together either using crisp (non-fuzzy) or fuzzy techniques.
  b. Association learning—identifies common groups of items that occur frequently together or, in more complex examples, if data item A, data item B will occur with definite probability.

(c) Semi-supervised learning—learning which occurs when the output variable is identified for only a portion of examples.
(d) Regression—prediction of a real-valued output variable, which includes particular examples of forecasting future values within a time series based on recent or past values (Mikut and Reischl 2011).

Other tasks include:

(a) Data cleaning (removal of redundant values, approximating missing values, etc.).
(b) Data filtering (including smoothing of time series).
(c) Feature extraction—identifying characteristic from images, videos, graphs, etc. Feature identification includes the sub-tasks of segmentation and segment description for images and identifying values such as common structures in graphs.
(d) Feature transformation—features transformed through mathematical operations such as logarithms, dimension reduction through principal component analysis, factor analysis, or independent component analysis.
(e) Feature evaluation and selection: using techniques of artificial intelligence, notably the filter and wrapper methods.
(f) Calculation of similarities and identification of the most similar items in terms of features through the use of correlation analysis or k-nearest neighbour techniques.

(g) Model validation—validation accomplished through the techniques of bootstrapping, statistical relevance checks, and complexity procedures.
(h) Model optimisation—through the use of many different techniques, including genetic algorithms (Mikut and Reischl 2011).

These techniques, in themselves, utilise other techniques to accomplish their goal. These other techniques include fuzzy models, support vector machines, random forest, estimated probability density function, artificial neural networks, and rough sets (Mikut and Reischl 2011).

A quick categorisation of the frequency of these methods as found in data mining tools is as follows:

(a) Frequently found—tools that use classifiers obtained through estimated probability density function, statistical feature selection, relevance checks, and correlation analysis techniques.
(b) Commonly found—tools that use decision trees and artificial neural networking techniques and perform tasks of clustering, regression, data cleaning, feature extraction, data filtering, principle component analysis, factor analysis, calculation of similarities, model cross validation, statistical relevance checks, advanced feature assessment and choice.
(c) Less likely found—tools that use independent component analysis, complexity procedures, bootstrapping, support vector machines, Bayesian networks, and discrete rule techniques while performing the tasks of fuzzy classification, model fusion, association identification, and mining frequent item sets.
(d) Rare—tools that use random forest, fuzzy system learning, and rough set techniques while performing the task of model optimisation through genetic algorithms. Random forests are incorporated within the tools of Waffles Weka, and random forests. Fuzzy system learning is incorporated with See5, Knowledge Miner, and Gait-CD. Use of rough sets is integrated in Rosetta and Rseslibs tools while model optimisation through genetic algorithm is performed by KEEL, Adam, and D2K tools (Mikut and Reischl 2011).

6 Data Importation and Data Processes

One of the most significant roles for big data tools is the importation of data, from various sources to manipulate and analyse. Traditionally, most tools supported the importation of text or comma-delimited data files. SAS and IBM tools support a XML data-exchange standard, PMML. In addition, to aid the connection of these tools to heterogeneous databases, a set of standard interfaces, object linking embedding (OLE), were defined and incorporated into objects that served as an intermediary between these databases and the tools querying them via the Structured Query Language (SQL). These tools included those produced by SAP, SAS, SPPS, and Oracle. However, besides common standards for data exchange, most tool have their own proprietary data formats, such as the Attribute-Relation File Format for the Weka tool (WEKA standard) (Mikut and Reischl 2011).

Other than the exchange of data, some data mining tools provide advanced aspects including data warehousing and Knowledge Discovery in Databases (KDD) procedures. A KDD is a procedure of identifying the most beneficial knowledge from a large cluster of data. A data warehouse could be defined as a storehouse of integrated data that is focused by subject and varied by time that is used to lead decisions by management. An example of a data warehouse might be the purchases at a grocery store over a given year that might yield information as to how much of a select product is selling and when its peak sales period is in order to assist management in ensuring that they have enough of the product on hand when the peak period hits (such as snow shovels at the start of winter) (Top 15 Best Free Data Mining Tools: The Most Comprehensive List 2019).

7 Tools

In this chapter, we look at the most popular data mining tools, each with their particular characteristics and advantages:

1. Rapid Miner—an open-source tool that supplies an integrated environment for the methods of machine learning, deep learning, text mining, and predictive analysis. This tool is capable of serving multiple application domains such as education, machine learning, research, business, and application development. Based on a client/server model, Rapid Miner can serve as both in-house and within private/public cloud infrastructures. Furthermore, Rapid Miner comes with a number of template-base frameworks that can be deployed quickly with fewer errors than the traditional manual code-writing method. This tool is comprised of three modules, each with a different purpose. These modules are the following:

    a. Rapid Miner Studio—it designed for prototyping, workflow design, validation, et al.
    b. Rapid Miner Server—it designed to deploy the predictive data models developed in Rapid Miner Studio.
    c. Rapid Miner Radoop—it designed to directly implement processes in the big data Hadoop cluster in order to streamline predictive analysis.

2. Orange—this open-source tool is well-suited for data mining, machine learning, and data visualisation. Designed as a component-based tool with the components termed "widgets", various widgets focus on different functions from data-preprocessing, evaluation of different algorithms, predictive modelling, and data visualisation. An additional advantage of Orange is its ability to quickly format incoming data to a set pattern so that it can easily be utilised by the tool's various widgets.

3. Weka—it is open-source software that contains a GUI that allows navigation to all of its aspects such as machine learning, data analysis, predictive modelling, and visualisation. Data importation is via a flat file or through SQL databases.
4. KNIME—it is open-source software that tightly incorporates machine learning, data mining, and reporting functions together. Besides quick deployment and efficient scaling, it has an easy learning curve for users. It is commonly used in research of pharmaceuticals but it also employed, with excellent results, for financial and customer data analysis and business intelligence.
5. Sisense—it is proprietary software that is best used for business intelligence and reporting within an organisation. This tool allows the integration of data from different sources to build a common depository and it further refines data to produce rich, highly visula reports for every unit within an organisation. These reports may be in the format of pie charts, bar graphs, line charts, et al. depending on the need. These reports allow the drilling down of items within them to obtain a wider set of data. This tool is particularly designed for non-technical users with a drag-and-drop ability with widgets.
6. Apache Mahout—it is an open-source tool whose main goal is to assist in the development of algorithms, particularly machine learning. As algorithms are developed, they are incorporated into this tool's growing libraries. This tool is able to conduct mathematical procedures such as linear algebra and statistics and concentrates on classification, clustering and collaborative filtering.
7. Oracle Data Mining—it is proprietary software that provides a "drag-and-drop" interface for easy use while leveraging the advantages of an Oracle database. This tool contains excellent algorithms for prediction, regression, data classification, and specialised analytics. In turn, these algorithms enable their users to leverage their data in order to focus on their best customers, find cross-selling opportunities, perform more accurate predictions, identify fraud, and further analyse identified insights from the data.
8. Rattle—it is an open-source tool that is based on the R programming language which, subsequently, provides the statistical functionality of R. In addition to providing a GUI-based coding interface to develop and extend existing code, Rattle allows the viewing and editing of the data that it utilises.
9. DataMelt—it is an open-source tool that provides an interactive environment for data analysis and visualisation. This tool is often used by engineers, scientists and students in the domains of engineering, the natural science, and financial markets. The tool contains mathematical and scientific libraries that enable it to draw two- or three-dimensional plots with curve fitting. This tool is capable of being used to analyse large data volumes, statistical analysis, and data mining.
10. IBM Cognos—it is a proprietary suite of software tools that are composes of parts designed to meet particular organisational needs. These parts include the following:

    a. Cognos Connection—it is a web portal that collects and summarises data within reports/scoreboards

b. Query Studio—it holds queries that format data and produce diagrams from them
c. Report Studio—it creates management reports
d. Analysis Studio—it is able to manage large data volumes will extracting patterns that indicate trends
e. Event Studio—it provides notifications of events transpiring
f. Workspace Advanced—it provides an interface to develop customised documents.

11. SAS Data Mining—it is a proprietary tool that can process data from heterogeneous sources, possesses a distributed memory architecture for easy scalability, and provides a graphical user interface for less technical users. This tool is able to change data, mine it, and perform various statistical analysis on it.
12. TeraData—it is a proprietary tool that is focused on the business market with providing an enterprise data warehouse with data mining and analytical capability. This tool provides businesses with insights derived from their data such as customer preferences, sales, and product placement with the ability to differentiate between "hot" and "cold" data where "cold" data, less frequently used data, is placed in a slow storage section.
13. Board—it is a proprietary tool that focuses on analytics, corporate performance management, and business intelligence. This tool provides one of the most comprehensive graphical user interfaces among all these tools and it is used to control workflows, track performance planning, and conduct multi-dimensional analysis. The goal of this software is to assist organisations who wish to improve their decision making.
14. Dundas—it is a proprietary tool that provides quick insights from data, rapid integration of data, and unlimited data transformation patterns which can produce a range of tables, charts and graphs. This tool uses multi-dimensional analysis with a speciality in business critical decisions.
15. H2O—it is an open-source tool that performs big data analysis on data held in various cloud computing applications and environments (Top 15 Best Free Data Mining Tools: The Most Comprehensive List 2019).

8 Conclusion

In this chapter, the various users of data mining tools were described along with a set of tasks which these tools might perform in order to provide a context for the particular choice of a tool. A most current list of the most popular data mining tools was given, along with a short description of their capabilities and often their target market in terms of user and task.

9 Key Terms and Definitions

Data mining—the process of identifying patterns of useful information within a large dataset, either discrete or data streaming.

Proprietary—belonging to a particular company, with usage often restricted by license.

Open Source—software where the source code is freely available, which may be modified for one's particular purpose. Open source entails the use of software by anyone without needing to obtain a license first.

References

Al-Azmi, A. A. R. (2013). Data, text and web mining for business intelligence: a survey. *arXiv preprint* arXiv:1304.3563.

Mikut, R., & Reischl, M. (2011). Data mining tools. *Wiley Interdisciplinary Reviews: Data Mining and Knowledge Discovery, 1*(5), 431–443.

Top 15 Best Free Data Mining Tools: The Most Comprehensive List (2019). Available from https://www.softwaretestinghelp.com/data-mining-tools/.

Richard Millham is currently an Associate Professor at Durban University of Technology in Durban, South Africa. After thirteen years of industrial experience, he switched to academe and has worked at universities in Ghana, South Sudan, Scotland, and Bahamas. His research interests include software evolution, aspects of cloud computing with m-interaction, big data, data streaming, fog and edge analytics, and aspects of the Internet of Things. He is a Chartered Engineer (UK), a Chartered Engineer Assessor and Senior Member of IEEE.

The manufacturer's authorised representative in the EU is Springer Nature Customer Service Centre GmbH, Europaplatz 3, 69115 Heidelberg, Germany. If you have any concerns regarding our products, please contact ProductSafety@springernature.com

Printed and bound by CPI Group (UK) Ltd, Croydon, CR0 4YY
07/10/2026
02229168-0002